New Frontiers in Regional Science: Asian Perspectives

Volume 6

Editor in Chief
Yoshiro Higano, University of Tsukuba

Managing Editors
Makoto Tawada (General Managing Editor), Aichi Gakuin University
Kiyoko Hagihara, Bukkyo University
Lily Kiminami, Niigata University

Editorial Board
Sakai Yasuhiro (Advisor Chief Japan), Shiga University
Yasuhide Okuyama, University of Kitakyushu
Zheng Wang, Chinese Academy of Sciences
Yuzuru Miyata, Toyohashi University of Technology
Hiroyuki Shibusawa, Toyohashi University of Technology
Saburo Saito, Fukuoka University
Makoto Okamura, Hiroshima University
Moriki Hosoe, Kumamoto Gakuen University
Budy Prasetyo Resosudarmo, Crawford School of Public Policy, ANU
Shin-Kun Peng, Academia Sinica
Geoffrey John Dennis Hewings, University of Illinois
Euijune Kim, Seoul National University
Srijit Mishra, Indira Gandhi Institute of Development Research
Amitrajeet A. Batabyal, Rochester Institute of Technology
Yizhi Wang, Shanghai Academy of Social Sciences
Daniel Shefer, Technion - Israel Institute of Technology
Akira Kiminami, The University of Tokyo

Advisory Board
Peter Nijkamp (Chair, Ex Officio Member of Editorial Board), Free University Amsterdam
Rachel S. Franklin, Brown University
Mark D. Partridge, Ohio State University
Jacques Poot, University of Waikato
Aura Reggiani, University of Bologna

New Frontiers in Regional Science: Asian Perspectives

This series is a constellation of works by scholars in the field of regional science and in related disciplines specifically focusing on dynamism in Asia.

Asia is the most dynamic part of the world. Japan, Korea, Taiwan, and Singapore experienced rapid and miracle economic growth in the 1970s. Malaysia, Indonesia, and Thailand followed in the 1980s. China, India, and Vietnam are now rising countries in Asia and are even leading the world economy. Due to their rapid economic development and growth, Asian countries continue to face a variety of urgent issues including regional and institutional unbalanced growth, environmental problems, poverty amidst prosperity, an ageing society, the collapse of the bubble economy, and deflation, among others.

Asian countries are diversified as they have their own cultural, historical, and geographical as well as political conditions. Due to this fact, scholars specializing in regional science as an inter- and multidiscipline have taken leading roles in providing mitigating policy proposals based on robust interdisciplinary analysis of multifaceted regional issues and subjects in Asia. This series not only will present unique research results from Asia that are unfamiliar in other parts of the world because of language barriers, but also will publish advanced research results from those regions that have focused on regional and urban issues in Asia from different perspectives.

The series aims to expand the frontiers of regional science through diffusion of intrinsically developed and advanced modern regional science methodologies in Asia and other areas of the world. Readers will be inspired to realize that regional and urban issues in the world are so vast that their established methodologies still have space for development and refinement, and to understand the importance of the interdisciplinary and multidisciplinary approach that is inherent in regional science for analyzing and resolving urgent regional and urban issues in Asia.

Topics under consideration in this series include the theory of social cost and benefit analysis and criteria of public investments, socioeconomic vulnerability against disasters, food security and policy, agro-food systems in China, industrial clustering in Asia, comprehensive management of water environment and resources in a river basin, the international trade bloc and food security, migration and labor market in Asia, land policy and local property tax, information and communication technology planning, consumer "shop-around" movements, and regeneration of downtowns, among others.

More information about this series at http://www.springer.com/series/13039

Lily Kiminami • Toshihiko Nakamura
Editors

Food Security and Industrial Clustering in Northeast Asia

Editors
Lily Kiminami
Institute of Science and Technology
Niigata University
Niigata, Niigata, Japan

Toshihiko Nakamura
Economic Research Institute
 for Northeast Asia
Niigata, Japan

ISSN 2199-5974　　　　　ISSN 2199-5982　(electronic)
New Frontiers in Regional Science: Asian Perspectives
ISBN 978-4-431-55281-9　　ISBN 978-4-431-55282-6　(eBook)
DOI 10.1007/978-4-431-55282-6

Library of Congress Control Number: 2015944539

Springer Tokyo Heidelberg New York Dordrecht London
© Springer Japan 2016
This work is subject to copyright. All rights are reserved by the Publisher, whether the whole or part of the material is concerned, specifically the rights of translation, reprinting, reuse of illustrations, recitation, broadcasting, reproduction on microfilms or in any other physical way, and transmission or information storage and retrieval, electronic adaptation, computer software, or by similar or dissimilar methodology now known or hereafter developed.
The use of general descriptive names, registered names, trademarks, service marks, etc. in this publication does not imply, even in the absence of a specific statement, that such names are exempt from the relevant protective laws and regulations and therefore free for general use.
The publisher, the authors and the editors are safe to assume that the advice and information in this book are believed to be true and accurate at the date of publication. Neither the publisher nor the authors or the editors give a warranty, express or implied, with respect to the material contained herein or for any errors or omissions that may have been made.

Printed on acid-free paper

Springer Japan KK is part of Springer Science+Business Media (www.springer.com)

Preface

This book consists of sections translated into English, revised, or written anew, based on *Food Security and Industrial Clusters in Northeast Asia* (Agriculture and Forestry Statistics Publishing Inc., March 2011), with editors Lily Kiminami and Toshihiko Nakamura. The Japanese-language version won the 22nd Annual Japan Section of the Regional Science Association International (JSRSAI) Award for academic literary work in 2013, and that became one of the inspirations for an English-language version. On the occasion of the publication of this book, I would like to express my gratitude to Mr. Shinichi Kawabe of Agriculture and Forestry Statistics Publishing Inc. for his generous support.

Northeast Asia as referred to in this book encompasses Japan, the People's Republic of China, the Republic of Korea (ROK), the Democratic People's Republic of Korea (DPRK), Mongolia, and the Russian Federation. If these countries are combined, they have the potential to form an economic region to rival or surpass North America or the EU in size. Among the different countries, however, there are political antagonisms, economic disparities, and gaps in social systems, and they have not fully made the most of their economic potential as a region. Regional cooperation has for the most part not been attempted regarding how to ensure food security and how to promote the growth of the food industry. This book, alongside elucidating the current situation for the food industries of Japan, China, and the ROK, introduces case examples of food industry clusters in North America and Europe. My hope is that it will provide a reference point for considering the future food security in Northeast Asia.

Niigata, Japan   Toshihiko Nakamura

Acknowledgments

The production of this book has involved the cooperation of many people over a number of years. The project has received funding from ERINA (the Economic Research Institute for Northeast Asia), and we have benefited from the interaction of many current and former colleagues and friends.

This book is the fruit of team learning. For all the years that went into this work, the special team supported, distracted, and challenged me in all the ways one could hope for. As Archimedes said, "Give me a lever long enough … and single-handed I can move the world."

Lastly, I would like to acknowledge the support of our extended family of JSRSAI (the Japan Section of the Regional Science Association International), as the completion of this book has reminded me of the humility one must have in academic work.

Lily Kiminami

Contents

1 **Food Security and Collaborative Advantage: Scoping the Scene** 1
Lily Kiminami

Part I Food Security from the Asian Perspective

2 **Defining Japan's Food Security in East Asia** 21
Yonosuke Hara

3 **Current Position and Future Direction of Agriculture in Northeast Asia** 33
Shin-ichi Shogenji

4 **Analysis of China's Food Supply and Demand Balance and Food Security** 47
Yongfu Chen and Fengying Nie

5 **Korea's Food Security Schemes** 61
Jaehyeon Lee

Part II Food Clustering in Northeast Asia

6 **The Food System Based on Agriculture** 75
Osamu Saito

7 **The Network Structure of a Soybean Cluster in Hokkaido** 87
Teruya Morishima

8 **The Competitive Advantages of Green Tea Clusters in Japan** 95
Yuko Akune

9 **Food and Health-Related Industry Clustering in Niigata Prefecture** 111
Lily Kiminami and Shinichi Furuzawa

| 10 | Agricultural Industry Clusters in China | 129 |

Lily Kiminami and Akira Kiminami

| 11 | Industrial Agglomeration of the Food Industry in China: An Analysis of Data by Province | 141 |

Hironori Yagi

| 12 | The Agricultural Industrialization of China's Heilongjiang Province | 157 |

Jiao Jiang

| 13 | Agricultural Production and Related Business by Public Firms: A Case Study on Xinhua Farm, Heilongjiang | 167 |

Hironori Yagi and Yonghao Zhu

| 14 | Promotion Policies for Food Industry Cluster in Korea | 179 |

Byung-Oh Lee

| 15 | The Trends and Potential for Food Industry Clusters in Korea | 197 |

Jaehyeon Lee

| 16 | The Promotion of and Challenges for the Agricultural Senary Industrialization Policy in the Republic of Korea | 209 |

Youkyung Lee

Part III Food Clustering in EU and North America

| 17 | Cluster Initiatives in Eastern Poland: Good Practices in Agriculture and Food-Processing Industry | 227 |

Ewa Bojar, Matylda Bojar, and Wiktor Bojar

| 18 | Main Factors Affecting Food Industry Clustering in France | 241 |

Nejla Ben Arfa and Karine Daniel

| 19 | Industrial Cluster Analysis, Entrepreneurship and Regional Economic Development | 255 |

Roger R. Stough and Junbo Yu

Contributors

Yuko Akune Faculty of Economics and Business Administration, Reitaku University, Kashiwa, Japan

Nejla Ben Arfa Social Sciences Laboratory, Ecole Supérieure d'Agriculture ESA, Angers, France

Ewa Bojar Department of Economics and Management of Economy, Lublin University of Technology, Lublin, Poland

Matylda Bojar Department of Management, Lublin University of Technology, Lublin, Poland

Wiktor Bojar Department of Small Ruminants Breeding and Agricultural Advisory, University of Life Sciences in Lublin, Lublin, Poland

Yongfu Chen College of Economics and Management, China Agricultural University, Beijing, China

Karine Daniel Social Sciences Laboratory, Ecole Supérieure d'Agriculture ESA, Angers, France

Shinichi Furuzawa Institute of Science and Technology, Niigata University, Niigata, Niigata, Japan

Yonosuke Hara National Graduate Institute for Policy Studies (GRIPS), Tokyo, Japan

Jiao Jiang Heilongjiang Academy of Agricultural Sciences, Harbin, China

Akira Kiminami Department of Agricultural and Resource Economics, Graduate School of Agricultural and Life Sciences, The University of Tokyo, Tokyo, Japan

Lily Kiminami Institute of Science and Technology, Niigata University, Niigata, Niigata, Japan

Byung-Oh Lee Kangwon National University, Chuncheon, Republic of Korea

Jaehyeon Lee Faculty of Agriculture, Kagoshima University, Kagoshima, Japan

Youkyung Lee Department of International Development Studies, College of Bioresource Sciences Nihon University, Fujisawa, Japan

Teruya Morishima Farm Management Division, National Agriculture and Food Research Organization (NARO), Agricultural Research Center, Tsukuba, Japan

Fengying Nie International Information Division, Agricultural Information Institute, Chinese Academy of Agricultural Sciences, Beijing, China

Osamu Saito Food and Resource Economics Course, Chiba University, Chiba, Japan

Shin-ichi Shogenji Graduate School of Bioagricultural Sciences, Nagoya University, Nagoya, Japan

Roger R. Stough George Mason University, Fairfax, VA, USA

Hironori Yagi Department of Agricultural and Resource Economics, Graduate School of Agricultural and Life Sciences, The University of Tokyo, Tokyo, Japan

Junbo Yu School of Administration, Jilian University, Jilian, China

Yonghao Zhu Faculty of Economics and Business Administration, Fukushima University, Fukushima City, Japan

Chapter 1
Food Security and Collaborative Advantage: Scoping the Scene

Lily Kiminami

Abstract Generally, the food we eat today has to travel a long distance "from farm to table". As for the agents connected with food, the roles that the distribution and processing industries play are growing, in addition to those of farmers and consumers. Moreover, since food production is based on biological production and located in rural areas, it is strongly subject to the influence of the natural environment, and its relationship with the social and economic background of farm villages is also important. Therefore, a perspective on the "food system", which covers problems involving foodstuffs such as the flow of food and the environment surrounding it, is needed.

On the other hand, since the issues of food security are conventionally regarded as problems in a single country (one sector) or problems at a certain stage of development, such a viewpoint has become an obstacle to WTO negotiations on agriculture, which makes the liberalization of global trade more difficult. Therefore, it is necessary to apply a framework that ranges across different areas of a country, different stages of development and relationships of interdependence among countries, and to cover these problems from an international perspective.

Additionally, in recent years, studies on industrial clusters have made significant achievements in regional development for seeking new growth strategies based on collaborative advantage which is a paradigm shift from the comparative and competitive advantages of conventional economics. The policy design for improving the capability of collaboration among entities and accepting the diversified and pluralistic nature of entities is an urgent necessity for solving the problem of food security and realizing the sustainability of regional development as well.

Keywords Food security • Food system • Collaborative advantage • Sustainability of regional development • Innovation

---

L. Kiminami (✉)
Institute of Science and Technology, Niigata University, Niigata, Niigata, Japan
e-mail: kiminami@agr.niigata-u.ac.jp

1 The Perspective on the Food System

Food is a general term including agricultural products, such as grains, vegetables, livestock products, fruits, oils and fats, sugars, marine products, and related processed goods. Generally, the food we eat today has to travel a long distance "from farm to table". As for the agents connected with food, the roles that the distribution and processing industries play are growing, in addition to those of farmers and consumers. Moreover, since food production is based on biological production and located in rural areas, it is strongly subject to the influence of the natural environment, and its relationship with the social and economic background of farm villages is also important. On the other hand, due to the influence of history and culture, etc., food consumption, in addition to being concerned with life, has relationships with various environments beyond the consumption of common goods. Furthermore, the problematic food-related domain is further expanded by the development of biotechnology, the preservation of biodiversity, and the promotion of the utilization of biomass, etc. Therefore, a perspective on the "food system", which covers problems involving foodstuffs such as the flow of food and the environment surrounding it, is needed (Kiminami 2009).

Since food problems are conventionally regarded as problems in a single country (one sector) or problems at a certain stage of development, such a viewpoint has become an obstacle to WTO negotiations on agriculture, which makes the liberalization of global trade more difficult. Therefore, when considering today's food problems, it is necessary to apply a framework that ranges across different areas of a country, different stages of development and relationships of interdependence among countries, and to cover these problems from an international perspective, rather than in one country (area) or one sector (field). Meanwhile "being international" in this case means that not only the supplier or source of the food are global, but also the local and global food economies are linked to each other.

2 Global Food Security and the Strategy of Collaborative Advantage

Here, we consider food security as a problem of the whole globe, comprising both the developed and developing countries, including newly emerging markets (see Fig. 1.1). Initially, developed countries are assumed to have met the minimum level of food security, but developing countries have not yet done so (point A). Along with the growth of the economy in developing countries, the level of food security in these countries is assumed to be improving. Under the same rationale, if there is no cooperative relationship on food security between developing and developed countries, there is a possibility that this may reduce the level of food security in developed countries (point A', or a shift to point C). However, if both developed and developing countries cooperate and compete with each other (converting to a complementary relationship via trade-offs), it is possible that the level of food security for both parties can be improved (point D).

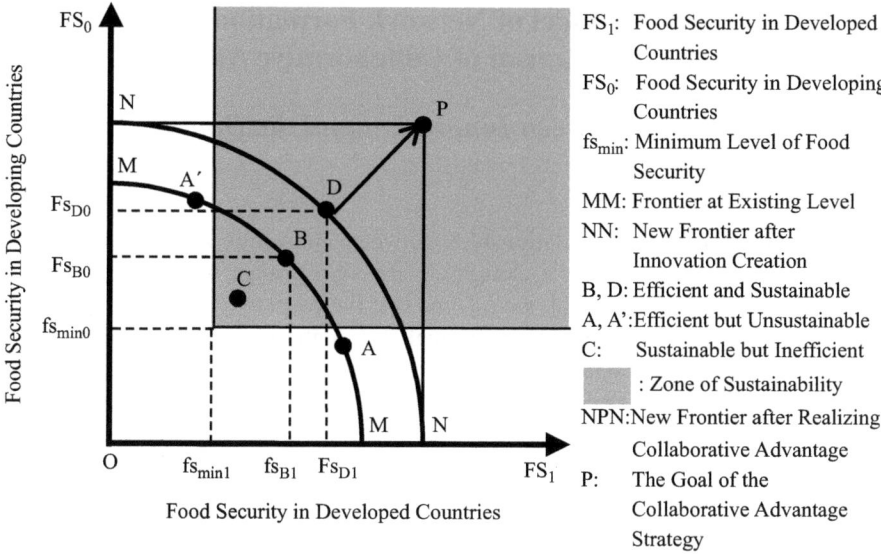

Fig. 1.1 Food security and collaborative advantage (Source: Kiminami (2011a, p. 13))

Additionally, in recent years, studies on industrial clusters have made significant achievements in regional development for seeking new growth strategies (Porter 1998).[1] However, the cluster strategy is based on collaborative advantage (Huxham 1996) which is a paradigm shift from the comparative and competitive advantages of conventional economics.

With a strategy of collaborative advantage, greater results can be obtained by strategically forming a network and collaboration between companies and regions, which contribute to sustainable development in particular. Therefore, Fig. 1.1 shows an ideal state for food security (point P), to be realized by innovation creation through cluster formation, which is a definite breakthrough in accordance with the principle of collaborative advantage. It could be a powerful strategy for the sustainable development of industries, regions and food security amid today's globalization. Furthermore, such a way of thinking advances the conventional discussion for food security oriented toward the food self-sufficiency ratio.

Moreover, existing empirical research shows that local food industrial clusters are also being formed in Northeast Asia, such as in Japan, China and South Korea. In addition, the food industry in Northeast Asia is strengthening the relationship among local clusters across national borders against the backdrop of geographical proximity, varied natural conditions, and the social and economic environments.

---

[1] Porter (2000) defined clusters as "geographic concentrations of interconnected companies, specialized suppliers, service providers, firms in related industries, and associated institutions (e.g., universities, standards agencies, trade associations) in a particular field that compete but also cooperate".

3 The Theoretical Model of Network Formation: The Incentive Mechanism of Collaborative Advantage

3.1 Relationships Between Innovation and the Quality (Price) of Products

Here we shall focus on the relationships between innovation and the quality and price of goods in the course of examining the significance of cluster formation through business alliances. First, we assume that there are two types of product differentiation, namely horizontal and vertical differentiation, and second, that the market is characterized by monopolistic competition in which there are many buyers and sellers, and price is determined by markups added to production costs. Such a market structure can be indicatively considered as a transitional stage from growth to maturation, rather than a nascent stage.

Figure 1.2a shows the typical economic benefits associated with ordinary innovation in companies. The vertical axis shows production costs/prices, and the horizontal axis represents quality, where G_A, G_B, G_C, G_D, and G_E are goods or services produced and sold by companies A, B, C, D, and E, respectively. WTP (willingness-to-pay) is a curve that indicates the price at which consumers are willing to pay for the quality of each product. For example, among the five types of companies, G_B represents the quality and cost of goods or services produced by company B at a middle level, and are sold at a middle-level price as well.

In addition, we assume here that the sources of differentiation of products and services mainly come from the knowledge and technologies held independently by each company which cannot be easily imitated. However, there are numerous other companies of the same type for each of companies A, B, C, D, and E, which results

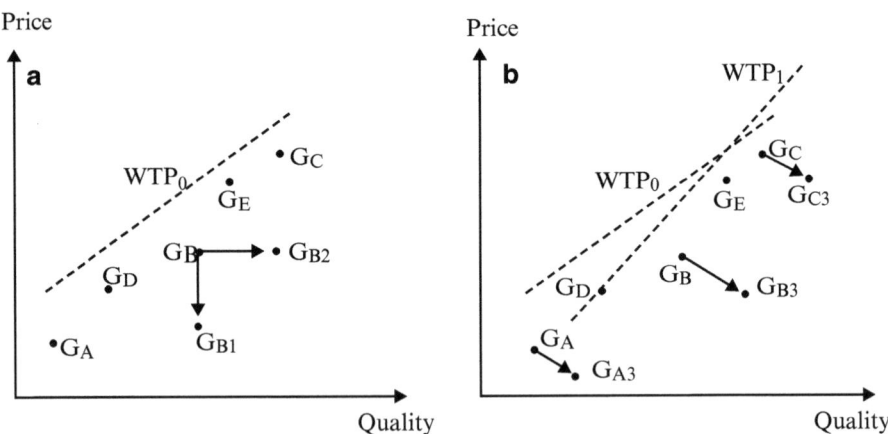

Fig. 1.2 Innovation and industrial clusters. (**a**) typical economic benefits, (**b**) production costs all go down, while the quality goes up (Source: Furuzawa and Kiminami (2011) with reference to Swann (2009, p. 53))

in horizontal differentiation. Furthermore, the vertical difference between WTP_0 and G_B represents the consumer surplus. Because the price for each company is equal to the costs of production plus a markup, there is no excess profit generated in this stage. Hence, each company adopts a strategy to maximize profits through innovation.

Here, suppose that company B succeeds in innovating. First, as the result of a process innovation, point G_B in Fig. 1.2a shifts downward to point G_{B1}, which is considered a cost-reduction innovation at a lower price level with a constant quality. In contrast, as the result of a product innovation, point G_B shifts right to point G_{B2}, which is considered a quality-enhancing innovation since quality goes up while production costs keep constant. Excess profits are generated and the market share of each company changes as the innovations are realized. Since each product can be replaced by others in either of the cases above (i.e., where G_B shifts to G_{B1} or G_B shifts to G_{B2}), the markets for products G_A, G_C, G_D, and G_E would shrink, and that for product G_B would grow. However, market shrinkages for G_C and G_E in the former case, and for G_A and G_D in the latter case, are respectively relatively large.

If three companies such as A, B, and C form a cluster through a business alliance and each company simultaneously realizes both process innovation and product innovation as shown in Fig. 1.2b, the production costs of their products G_A, G_B, and G_C all go down, while the quality goes up (shifts to G_{A3}, G_{B3}, G_{C3}) as a result of the spillover effects. Meanwhile, as the market becomes saturated in terms of quantitative size and the rise in living standards, a higher quality of goods and services would be demanded by consumers, which shifts the WTP curve from WTP_0 to WTP_1. Furthermore, if the entities that make up the cluster are able to establish win-win relationships, the competitiveness of the entire cluster would be improved which would lead to a concurrent increase in consumer surplus. In contrast, the demand for the products produced by companies D and E, which are not in the cluster, declines.

3.2 The Policy Implications of Network Formation

In the following section, the profit-maximization behavior of enterprises will be theoretically formulated by considering business alliances as the network formation, and the influence of policy on the network formation of enterprises will be taken into account as well.[2]

The profit of an enterprise, π, can be stated as Eq. 1.1, where, P is the price of the product, $F(L, K, N)$ is the production function, and A is the technical level of the enterprise. L, K and N are the inputs of labor, capital and network formation (stock), respectively. We assume the prices of each factor to be w, r, and b, in which w and r are decided by the market. On the other hand, the cost of network formation contains the cost of searching for partners, consensus building to form a network and maintaining the network, which is difficult to identify directly. Once the network is

---
[2] Please see Furuzawa and Kiminami (2011).

formed, costs will be generated and the enterprises in the network will be drained if the effectiveness of the network cannot be produced. Furthermore, the effectiveness of a network is dependent on the amount and the quality of information about business partners, the content of transactions and the business alliance, and the predictions regarding partners' behavior.[3]

$$\pi = P(N) \cdot F(L, K, N, A) - (wL + rK + bN) \qquad (1.1)$$

The first order condition of profit maximization can be represented as follows:

$$MPL = \partial F / \partial L = w / P \qquad (1.2)$$

$$MPK = \partial F / \partial K = r / P \qquad (1.3)$$

The equilibrium condition of network formation can be expressed using the concept of marginal revenue as follows:

$$MRN = (\partial F / \partial N) \cdot P + (\partial P / \partial N) \cdot F = b \qquad (1.4)$$

It is safe to say that success in development of new products through marketing and business alliances will cause a rise in the sales price per unit[4] and the level of network formation will be decided at the crossing point of the marginal revenue curve (the effects of the network on the increase in production + the increase in unit price) and the cost of network formation (Fig. 1.3). Additionally, the promotion of business alliances through subsidies is considered as the decrease in the cost of network formation from b to $b-s$. In other words, the effect of policy is considered as a part of the externality of developed technology and business model spread as public knowledge after the business alliance succeeds. Moreover, the economy of agglomeration is realized through geographic concentration, etc., which reduces the transaction costs of enterprises that lead to a decline in the cost of network formation, and the spillover of knowledge leads to the improvement of quality (price increase).

However, the effect of the spillover of knowledge formation and the cost of the network is not only affected by the level but also by the structure of networks (the proximity of the network). As for the structure of networks, the formation of small-world networks as opposed to random networks, would result in lower costs associated with network formation, in addition to greater economic benefits from higher prices. The policy implications of network formation are that the quantitative and qualitative differences in networks not only have consequences for the business

---

[3] For instance, the enactment of a commercial code improves the incentive for business contacts to abide by the rules, and improves the effectiveness of dealings. In addition, the cost of network formation for an enterprise is decreased via the facilitation of the prediction of the business contacts' actions.

[4] Moreover, it is necessary to consider the relationship with the content of the contract, including the purpose and the distribution of profits, when thinking about the reality of corporate behavior.

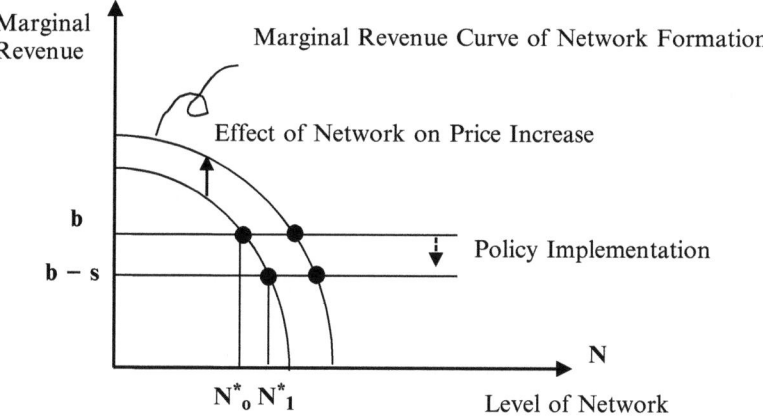

Fig. 1.3 Policy and optimal point of network formation (Source: Revised based on Kiminami (2011b, p. 248))

performance of corporate entities, but also have the potential to bring about differences in regional development (Kiminami et al. 2010, p. 468).

Therefore, collaborative advantage is understood as a strategy of the maximization of effects through the minimization of the cost (transaction costs in the broad sense) of governance and maximizing the benefits of "knowledge creation potential". The cost of governance here can be summarized from the perspectives of transaction costs theory, agency theory and ownership theory as follows: first, the transaction cost in the narrow sense of consensus building for cluster formation; second the cost of enforceable rules for suppression of the principal-agent problem based on information asymmetry, such as moral hazard; and third, the cost of preventing the problem of a free ride through the operation and regulation of knowledge ownership and promoting spillover and knowledge creation.

4 Selective Review of Research on Cluster Formation and Regional Innovation

4.1 Research on Agricultural Clusters

Relatively little research has been done on clusters which relate specifically to the agricultural and food sectors. There are some notable examples of studies, including Lagnevik et al. (2003), Hauknes (2001), Bertolini and Giovannetti (2003), and the European Monitoring Centre on Change (2006). However, the body of research into the role that agricultural cluster formation plays in the economic development of agriculture and rural communities is still small.

4.2 Research on Cluster Formation and Regional Development

Porter (2003) found that clusters affect regional employment, wages, and innovation levels to a great degree, which strengthened the theoretical foundations for the notion of regional development through cluster formation. There are also a number of studies that examine the role of clusters in regional development. For example, Porter et al. (2004) analyzed clusters from the perspective of regional development in economically advanced countries, while Ketels et al. (2006), OECD (2005), Bojar and Olesiński (2007), Kuchiki and Tsuji (2005), Otsuka and Sonobe (2006), Kuchiki (2007), and Ding (2007) took the same perspective but studied developing countries and Eastern European countries. Another pertinent example is Puppim de Oliveira (2008), the thesis of that the key to innovation and dynamic economic development is "social upgrading" among SMEs and clusters in developing countries.

4.3 Research on Cluster Initiatives

Most conventional studies on industrial clusters do not explicitly state who or what drives cluster formation and translates the benefits of clusters into real economic results, nor the way in which such results are achieved. The development and competitiveness of clusters rely largely on organized campaigns called cluster initiatives (CIs), which seek to advance precisely for the goals of those organizations that have ties to the relevant companies, the government, and research institutions within the region. Given this trend, one method effective for analyzing clusters is to focus not on the clusters themselves so much as on these cluster initiatives, taking into consideration the broad spectrum of circumstances in which these clusters were formed, and assessing them within the framework of cluster initiative models, i.e., examining what sort of influence factors such as initial conditions, purposes, and processes have had on the results. In the field of CI research, once the initial conditions, purposes, processes, and results are known, comparative analyses can be performed which measure the effects of those initial conditions, purposes, and processes on the results. In fact, a large-scale international CI survey project is currently underway by the Global Cluster Initiative Survey (GCIS), and as part of the project studies by Sölvell et al. (2003) and Ketels et al. (2006) have clearly identified the importance of CIs in cluster formation.

4.4 Research on the Economic Effects of Clusters

Most existing research on the economic effects of clusters tends to focus on industrial agglomeration. Theories of economic growth in recent years have recognized that innovation is essential for sustained growth, that knowledge (or technology) spillover is the root of such innovation, and that industrial agglomeration contributes to economic growth by promoting spillover. Therefore, it is possible to think of an economy of agglomeration as a manifestation of a dynamic external economy. The external economic effects of what we know as "spillovers" were noted by Marshall (1890), formulated by Arrow (1962), and applied to endogenous economic growth models by Romer (1986). According to Glaeser et al. (1992), dynamic external economies can be classified according to differences in the locations and market climates in which spillovers occur. Spillover can occur within a single industry or among different industries. It can also be spurred on by monopolistic/oligopolistic markets on one hand, and competitive markets on the other. Marshall (1890), Arrow (1962), Romer (1986), and Porter (1990) focused on spillover within a single industry, and theorized that regional specialization in an industry contributes to economic growth. Marshall (1890), Arrow (1962), and Romer (1986) speculated that monopolistic market structures facilitate spillover, but Porter (1990), in contrast, holds that competition promotes spillover. Jacobs (1969), on the other hand, places an emphasis on spillover among different industries, asserting that the agglomeration of diverse industries drives the creation of ideas and facilitates innovation. By extension, industry diversification contributes to economic growth.

It should be noted that because it is practically impossible to measure spillover directly, most empirical research relies on analyses of the relationship between industrial agglomeration and economic performance. For example, Glaeser et al. (1992) use an employment growth approach to perform an empirical analysis that explicitly adopts the notion of dynamic external economies through industrial agglomeration, while Beeson (1987), Dekle (2002), and Henderson (2003) take total factor productivity approaches to the same. However, as McCann (2008) points out, there are problems inherent in economic analyses of industrial agglomerations posed by understandings of the agglomerations themselves. Specifically, industrial agglomerations may make it easier for spillovers to occur, but that does not mean that spillovers necessarily *do* occur because of industrial agglomeration. When examining agglomeration economies, it is at least necessary to know the extent of business collaborations. However, exhaustively ascertaining all business collaborations among companies would require a study of massive proportions involving micro-level data. As is clear from the above, many issues remain for the study of the economic effects of clusters.

4.5 Research on Regional Innovation Strategy and Innovation System

Innovation is the buzzword of the times, although misunderstanding of this concept is also prevalent. Rather than technological progress, innovation is "creative destruction". Through paying attention to the contemporary context in which management finds itself and returning to the classic definition of innovation, Christensen (1997) rediscovered the essence of innovation using the concept of "destructive innovation", which was first focused on by Schumpeter (1934).[5]

Since the beginning of the 1990s, the theory of regional innovation has spread as policy in North America and the clusters introduced by Porter (1998), based on the theory of competitive advantage accumulated through knowledge in theoretical, empirical and policy terms (Cooke et al. 2004; Lundvall 1992). However, the following two problems have been pointed out by Matsubara (2013, pp. 11, 22), such as the insufficiency in the theoretical consideration of seeing the picture for the region and the knowledge spillover, and a lack of detailed empirical analysis about the actual condition of regional innovation systems. Moreover in Tödtling and Trippl (2005), the obstacles in regional innovation are classified into the surrounding area, the long-term industrial area, and the metropolitan area. Furthermore, the obstacles for the regions in the studies having the two aspects of the "surrounding area" and the "long-term industrial area" are the shortage (thinness of an organization) of resources, such as innovative companies, talented people, and research institutions, etc., and the fixation (lock-in) of relationships among industry, academia and government (Matsubara 2013, pp. 22–23).

Therefore, if innovation in a given region is regarded as a process of knowledge creation and application of the interaction of related entities, analysis of the relationship among innovation, the cognitive characteristics of corporations and the knowledge spillover in the process of cluster formation is considered to be important.

5 Sensitive Issues for Food Security in Northeast Asia

5.1 Factor Analysis on Food and Agricultural Trade Among China, Japan and S. Korea[6]

Questions over changes in the balance of trade between countries tend to arise in debates over the issues of food and agricultural trade. From a practical standpoint, however, it is reasonable to conduct the analysis of agricultural trade from the viewpoint of linking the agricultural sector of a given country with the total sectors of each country. Here, we suppose the change in the food and agricultural exports from Country A to Country B consists of three factors, namely: (i) the change in the importance of food and agricultural exports of Country A; (ii) the change in the exports of the total sectors of Country A; and (iii) the change in the importance of Country B as the destination for exports from Country A.

The value of sector i's exports from Country A to Country B (X_{abi}) is expressed as follows:

---

[5] See also Schumpeter (1950) for the classification of innovation.
[6] This section is revised based on Kiminami (2010).

1 Food Security and Collaborative Advantage: Scoping the Scene

$$X_{abi} = X_{abi}/X_{awi} \times X_{awi}/X_{aw} \times X_{aw} \qquad (1.5)$$

where X is the value of exports; a is country A; b is country B; w is the world; and i is the ith sector.

Equation 1.5 can be converted to Eq. 1.6 as follows:

$$G(X_{abi}) = G(X_{abi}/X_{awi}) + G(X_{awi}/X_{aw}) + G(X_{aw}) \qquad (1.6)$$

where $G(\cdot)$ = growth rate function; $G(X_{abi})$ = growth rate of the value of sector i's exports from country A to country B; $G(X_{abi}/X_{awi})$ = growth rate of the share of country B in the value of sector i's exports from country A; $G(X_{awi}/X_{aw})$ = growth rate of the share of sector i in the value of exports from country A; $G(X_{aw})$ = growth rate of the value of exports from Country A.

In the following sections, the international trade in food and agricultural products among Japan, China and S. Korea since 1985 will be analyzed using the above factor decomposition.

5.1.1 Food and Agricultural Exports from Japan to China and S. Korea

The most important factor for explaining the rise in food and agricultural exports from Japan to China before 1990 is the growth of Japan's total exports. Since 1990, as China's economy and population grew, so did its importance as a destination for food and agricultural exports, eventually becoming the most important among the factors. As a result, the Japanese food sector became a superior export sector compared with other sectors after 1995; particularly in the years after 2000, this trend had been continually strengthened up to 2005. Since 2005, China's priority as an importer for Japanese food and agricultural products has declined (Table 1.1).

Likewise, the growth of food and agricultural exports from Japan to S. Korea had been spurred by the importance of Korea as an importer up to 2000. Although S. Korea's priority as an importer for Japanese food and agricultural products has declined since then, the initial growth of food exports to the country had been caused by both the growth of total exports and the increase in the share of food and agricultural products in the total exports of Japan. However, the importance as an export market for Japanese food and agricultural products declined after 2005 (Table 1.2).

5.1.2 Food and Agricultural Exports from China to Japan and S. Korea

The growth of food and agricultural exports from China to Japan stems primarily from the growth of China's total exports. On the other hand, the superiority of the food and agricultural sectors, in terms of exports, and the priority of Japan as an importer declined somewhat in this period (Table 1.3).

Table 1.1 Decomposition of the growth of food exports from Japan to China. All figures are percentages

		Factor		
	Growth rate of export value	Importance as export market	Importance as export sector	Advantage of total exports toward world
Period	$G(X_{abi})$	$G(X_{abi}/X_{awi})$	$G(X_{awi}/X_{aw})$	$G(X_{aw})$
1985–1990	5.6	1.8	−6.0	10.3
		(29.4)	*(−97.6)*	*(168.2)*
1990–1995	28.8	26.0	−6.3	9.1
		(90.2)	*(−21.8)*	*(31.5)*
1995–2000	9.1	7.2	0.2	1.6
		(80.0)	*(2.3)*	*(17.7)*
2000–2005	20.4	12.7	2.3	4.4
		(65.4)	*(11.8)*	*(22.7)*
2005–2010	6.3	−3.4	4.5	5.3
		(−52.9)	*(70.5)*	*(82.4)*

Source: UN COMTRADE (United Nations Commodity Trade Statistics Database, retrieved 11 July 2014) http://comtrade.un.org/, SITC Revision No. 2 [Section 0. Food and Live Animals]
Note: Figures in parentheses are each factor's contribution to "$G(X_{abi}/X_{awi}) + G(X_{awi}/X_{aw}) + G(X_{aw})$"

Table 1.2 Decomposition of the growth of food exports from Japan to S. Korea. All figures are percentages

		Factor		
	Growth rate of export value	Importance as export market	Importance as export sector	Advantage of total exports toward world
Period	$G(X_{abi})$	$G(X_{abi}/X_{awi})$	$G(X_{awi}/X_{aw})$	$G(X_{aw})$
1985–1990	26.9	22.3	−6.0	10.3
		(83.8)	*(−22.4)*	*(38.6)*
1990–1995	14.8	12.3	−6.3	9.1
		(81.4)	*(−41.5)*	*(60.1)*
1995–2000	18.3	16.2	0.2	1.6
		(90.0)	*(1.2)*	*(8.8)*
2000–2005	7.7	0.8	2.3	4.4
		(10.4)	*(30.6)*	*(58.9)*
2005–2010	5.6	−4.0	4.5	5.3
		(−69.9)	*(78.3)*	*(91.6)*

Source: As Table 1.1
Note: As Table 1.1

Meanwhile, food and agricultural exports from China to S. Korea increased rapidly in the 1990s, a phenomenon caused mainly by the increased priority of S. Korea as an importer. However, since 2000, with its decline in priority, food and agricultural exports from China to S. Korea have declined as well (Table 1.4).

Table 1.3 Decomposition of the growth of food exports from China to Japan. All figures are percentages

Period	Growth rate of export value $G(X_{abi})$	Factor		
		Importance as export market $G(X_{ab}/X_{awi})$	Importance as export sector $G(X_{aw}/X_{aw})$	Advantage of total exports toward world $G(X_{aw})$
1985–1990	99.1	65.0	1.1	19.4
		(76.0)	*(1.3)*	*(22.7)*
1990–1995	21.0	11.9	−9.3	19.1
		(54.8)	*(−42.6)*	***(87.8)***
1995–2000	4.1	−0.3	−5.9	10.9
		(−5.6)	*(−125.0)*	***(230.6)***
2000–2005	8.2	−4.1	−9.8	25.0
		(−37.1)	*(−87.6)*	***(224.7)***
2005–2010	2.8	−8.9	−2.4	15.7
		(−204.1)	*(−55.8)*	***(359.9)***

Source: As Table 1.1
Note: As Table 1.1

Table 1.4 Decomposition of the growth of food exports from China to S. Korea. All figures are percentages

Period	Growth rate of export value $G(X_{abi})$	Factor		
		Importance as export market $G(X_{ab}/X_{awi})$	Importance as export sector $G(X_{aw}/X_{aw})$	Advantage of total exports toward world $G(X_{aw})$
1985–1990	–	–	–	–
	–	–	–	–
1990–1995	7.7	−0.3	−9.3	19.1
		(−3.3)	*(−97.4)*	***(200.7)***
1995–2000	27.4	22.1	−5.9	10.9
		(81.6)	*(−21.8)*	*(40.2)*
2000–2005	12.8	0.0	−9.8	25.0
		(−0.2)	*(−64.0)*	***(164.2)***
2005–2010	3.3	−8.5	−2.4	15.7
		(−176.7)	*(−50.8)*	***(327.5)***

Source: As Table 1.1
Note: As Table 1.1

5.1.3 Food and Agricultural Exports from Korea to Japan and China

Food and agricultural exports from S. Korea to Japan have been falling since 1995, a phenomenon attributed chiefly to the decline in Japan's priority as an importer and the decline in the competitiveness of Korea's food sector in terms of the nation's total exports (Table 1.5). Despite this decline in competitiveness, food and agricultural exports from S. Korea to China have been on the rise, primarily due to the increase in Korea's total exports and the increase in China's priority as an importer (Table 1.6).

Table 1.5 Decomposition of the growth of food exports from S. Korea to Japan. All figures are percentages

Period	Growth rate of export value $G(X_{abi})$	Factor		
		Importance as export market $G(X_{abi}/X_{awi})$	Importance as export sector $G(X_{awi}/X_{aw})$	Advantage of total exports toward world $G(X_{aw})$
1985–1990	11.5	−0.5	−3.7	16.5
		(−4.5)	(−30.6)	*(135.1)*
1990–1995	4.5	−1.0	−7.4	14.0
		(−18.8)	(−131.8)	*(250.6)*
1995–2000	−1.4	0.5	−8.0	6.6
		(−60.2)	*(923.8)*	*(−763.6)*
2000–2005	−6.3	−6.8	−9.0	10.5
		(128.1)	*(170.0)*	*(−198.1)*
2005–2010	4.4	−4.8	−0.6	10.4
		(−96.6)	(−12.8)	*(209.4)*

Source: As Table 1.1
Note: As Table 1.1

Table 1.6 Decomposition of the growth of food exports from S. Korea to China. All figures are percentages

Period	Growth rate of export value $G(X_{abi})$	Factor		
		Importance as export market $G(X_{abi}/X_{awi})$	Importance as export sector $G(X_{awi}/X_{aw})$	Advantage of total exports toward world $G(X_{aw})$
1985–1990	–	–	–	–
	–	–	–	–
1990–1995	13.4	7.4	−7.4	14.0
		(52.6)	(−52.6)	*(99.9)*
1995–2000	6.4	8.5	−8.0	6.6
		(119.6)	(−113.1)	(93.5)
2000–2005	16.4	15.8	−9.0	10.5
		(91.3)	(−52.3)	*(61.0)*
2005–2010	17.1	6.7	−0.6	10.4
		(40.8)	(−3.9)	*(63.1)*

Source: As Table 1.1
Note: As Table 1.1

According to the above-mentioned analyses, the interdependence of international food and agricultural trade among Japan, China and S. Korea can be summarized as follows. First, since 1985, food and agricultural exports from China to Japan and S. Korea have increased consistently, although this rise was caused by different factors. In the 1990s, both Japan and S. Korea were important export markets for Chinese food and agricultural products. However, their relative importance has fallen after 2000. Second, for Japan and S. Korea, the relative importance of China as an importer is rising whereas the importance of Japan for S. Korea and that of S. Korea

Table 1.7 Basic structure of networks

		1985	1990	1995	2000	2005
Distance index	Diameter	3	4	4	4	4
	Average distance	1.605	1.732	1.847	1.842	1.947
Cohesion index	Degree of density	0.432	0.374	0.337	0.326	0.295
	Transitivity	0.602	0.564	0.564	0.507	0.504

Source: Kiminami (2009), Kiminami and Furuzawa (2014)
Notes: Calculated from *Asian International Input–Output Table*, editions for 1985, 1990, 1995, 2000, and 2005, Institute of Developing Economies using the Simple Network Analysis Tool (software)

for Japan as a food importer are declining. In other words, China's influence in Northeast Asia as a major food consumer, not only as a food supplier, is growing.

5.2 Regional Interdependence in the Agriculture and Food Manufacturing Sectors in Northeast Asia

By using the Asian International Input–Output Table, we take a look at the changes in the network of agricultural and food manufacturing industries from 1985 to 2005 (Table 1.7). The results show that the bonding effect of the whole network is declining, while the average distance increased from 1.605 in 1985 to 1.947 in 2005, and in addition to the degree of density decreased from 0.432 to 0.295.

Furthermore, the change in network centeredness is summarized in Table 1.8. First, the centrality of the food manufacturing industry is higher than that of agriculture as a whole, and a network of agriculture and food manufacture centering on Japan was formed during the period of 1985–1990, and continued until 2000. However, the centrality of China's food manufacturing industry (AC008) grew from 2000 to 2005. The changes in the centrality of networks are considered as a reflection of the following three transactions: (i) intermediate inputs from Malaysia's agricultural sector to China's food manufacturing sector (from AM001–003 to AC008); (ii) intermediate inputs from Thailand's agricultural sector to China's food manufacturing sector (from AT001–003 to AC008); and (iii) intermediate inputs from China's food manufacturing sector to the U.S. food manufacturing sector (from AC008 to AU008). Furthermore, the driving forces of the transactions which strengthened the presence of China's food system in the region are considered to be the early harvest for the agricultural sector in the FTA with ASEAN in 2003 and China's accession to the WTO.[7]

---

[7] China and ASEAN signed the comprehensive framework agreement in November 2002 and the FTA started in full in July 2005 (ACFTA). In advance of the ACFTA, the customs duty on agricultural products of types 1–8 among the double figure HS codes was reduced to 10 % of the highest tax rate from 1 January 2004, as the early harvest measure. It was agreed to abolish tariffs gradually (Thailand carried this out in October 2003, and the Philippines in January 2006).
See JETRO Business Information Service Division (2012) and Kiminami and Furuzawa (2014) for more details.

Table 1.8 Centrality of the network

		1985		1990		1995		2000		2005	
		Degree	(Stand.)	Degree	(Stand.)	Degree	(Stand.)	Degree	(Stand.)	Degree	(Stand.)
Indonesia	AI001-003	4	0.211	6	0.316	3	0.158	4	0.211	3	0.158
	AI008	9	0.474	9	0.474	5	0.263	9	0.474	6	0.316
Malaysia	AI001-003	8	0.421	5	0.263	5	0.263	4	0.211	3	0.158
	AI008	12	0.632	7	0.368	10	0.526	10	0.526	7	0.368
Phillipines	AI001-003	5	0.263	3	0.158	2	0.105	3	0.158	4	0.211
	AI008	9	0.474	8	0.421	10	0.526	5	0.263	3	0.158
Singapore	AI001-003	10	0.526	11	0.579	10	0.526	14	0.737	9	0.474
	AI008	14	0.737	13	0.684	13	0.684	12	0.632	12	0.632
Thailand	AI001-003	7	0.368	5	0.263	1	0.053	3	0.158	3	0.158
	AI008	11	0.579	9	0.474	9	0.474	10	0.526	9	0.474
China	AI001-003	3	0.158	2	0.105	2	0.105	2	0.105	1	0.053
	AI008	8	0.421	7	0.368	8	0.421	7	0.368	10	0.526
Taiwan	AI001-003	4	0.211	1	0.053	1	0.053	1	0.053	1	0.053
	AI008	7	0.368	8	0.421	8	0.421	5	0.263	4	0.211
Korea(S.)	AI001-003	1	0.211	1	0.053	1	0.053	1	0.053	2	0.105
	AI008	12	0.632	10	0.526	9	0.474	6	0.316	6	0.316
Japan	AI001-003	2	0.105	2	0.105	2	0.105	2	0.105	4	0.211
	AI008	18	0.947	17	0.895	13	0.684	13	0.684	12	0.632
U.S.A.	AI001-003	5	0.263	5	0.263	6	0.316	4	0.211	2	0.105
	AI008	15	0.789	13	0.684	10	0.526	9	0.474	11	0.579
Centralization			0.573		0.579		0.386		0.456		0.374

Source: As Table 1.7
Notes: "001–003" is an aggregation of "Paddy", "Agricultural products" and "Livestock". "008" indicates "Food, beverages and tobacco". The top three countries/sectors are highlighted for each year

6 Ongoing as well as New Challenges

A common thread to the above analytical results has been picked out in this research that the issues of food security are not solely the problems of dealing agricultural and food products among other countries. Moreover, since the type of knowledge, the optimal timing and combination of networks in the clustering change dynamically, the policy design for improving the capability of collaboration among entities and accepting the diversified and pluralistic nature of entities in the network is an urgent necessity for solving the problem of food security in Northeast Asia and realizing a sustainable development of the region as well.

References

Arrow KJ (1962) The economic implications of learning by doing. Rev Econ Stud 29:155–173
Beeson PE (1987) Total factor productivity growth and agglomeration economies in manufacturing, 1959–73. J Reg Sci 27(2):183–199
Bertolini P, Giovannetti E (2003) The internationalisation of an agri-food cluster: a case study. Paper presented for the conference on clusters, industrial districts and firms: the challenges of globalization, Modena
Bojar E, Olesiński Z (eds) (2007) The emergence and development of clusters in Poland. Difin, Warsaw
Christensen CM (1997) The innovator's dilemma: when new technologies cause great firms to fail. Harvard Business School Press, Boston
Cooke P, Heidenreich M, Braczyk HJ (eds) (2004) Regional innovation systems, 2nd edn. Routledge, London
Dekle R (2002) Industrial concentration and regional growth: evidence from the prefectures. Rev Econ Stat 84(2):310–315
Ding K (2007) Domestic market-based industrial cluster development in modern China. IDE (Institute of Developing Economies) discussion papers 88
European Monitoring Centre on Change (2006) The food cluster in the Øresund region. European Foundation for the Improvement of Living and Working Conditions, Dublin
Furuzawa S, Kiminami L (2011) Theoretical and policy examinations concerning industrial clusters in Japan. In: ERSA2011, Barcelona, 30 Aug–3 Sept 2011. http://www-sre.wu.ac.at/ersa/ersaconfs/ersa11/e110830aFinal01850.pdf
Glaeser E, Kallal H, Scheinkman J, Shleifer J (1992) Growth in cities. J Polit Econ 100:1126–1152
Hauknes J (2001) Innovation styles in agro-food production in Norway. In: OECD (eds) Innovative clusters: drivers of national innovation systems. OECD Publishing, Paris, pp 157–178
Henderson JV (2003) Marshall's scale economy. J Urban Econ 53:1–28
Huxham C (1996) Creating collaborative advantage. Sage, London
Jacobs J (1969) The economy of cities. Vintage, New York
JETRO Business Information Service Division (2012) Goods trade agreement of ASEAN-China Free Trade Agreement (ACFTA) (January 2012) [in Japanese]
Ketels C, Lindqvist G, Sölvell Ö (2006) Cluster initiatives in developing and transition economies. Center for Strategy and Competitiveness, Stockholm
Kiminami L (2009) International food system. Agriculture and Forestry Statistics Publishing, Tokyo [in Japanese]
Kiminami L (2010) Agricultural trade within Northeast Asia: the prospects from Japan. J US-China Public Adm 7(5):17–29
Kiminami L (2011a) Food security and food policy. In: Kiminami L, Nakamura T (eds) Food security and industrial clusters in northeast Asia. Agriculture and Forestry Statistics Publishing, Tokyo, pp 1–22 [in Japanese]
Kiminami L (2011b) Concluding remarks: international industrial cluster and Niigata. In: Kiminami L, Nakamura T (eds) Food security and industrial clusters in northeast Asia. Agriculture and Forestry Statistics Publishing, Tokyo, pp 245–255 [in Japanese]
Kiminami L, Furuzawa S (2014) Dynamic changes in China's food system. Stud Reg Sci 44(1):41–62
Kiminami L, Kiminami A, Furuzawa S, Nakamura T, Zhu Y (2010) An analysis on the business alliances in food-related industries: international comparison between Niigata prefecture of Japan and Heilongjiang province of China. Stud Reg Sci 40(2):449–471 [in Japanese]
Kuchiki A (2007) The flowchart model of cluster policy: the automobile industry cluster in China. IDE (Institute of Developing Economies) discussion papers 100
Kuchiki A, Tsuji M (eds) (2005) Industrial clusters in Asia: analyses of their competition and cooperation. Palgrave Macmillan, New York

Lagnevik M, Sjöholm I, Lareke A, Östberg J (2003) The dynamics of innovation clusters: a study of the food industry. Edward Elgar, Cheltenham
Lundvall BÅ (ed) (1992) National systems of innovation. Pinter, London
Marshall A (1890) Principles of economics. Macmillan, New York
McCann P (2008) Agglomeration economics. In: Karlsson C (ed) Handbook of research on cluster theory. Edward Elgar, Cheltenham, pp 23–38
Matsubara H (2013) Industrial clusters and theory of regional innovation. In: Matsubara H (ed) Cluster policy and regional innovation in Japan. University of Tokyo Press, Tokyo, pp 3–25 [in Japanese]
OECD (2005) Business clusters: promoting enterprise in Central and Eastern Europe. OECD Publishing, Paris
Otsuka K, Sonobe T (2006) Strategy for cluster-based industrial development in developing countries. In: Ohno K, Fujimoto T (eds) Industrialization of developing countries: analyses by Japanese economists. National Graduate Institute of Policy Studies, Tokyo, pp 67–79
Porter M (1990) The competitive advantage of nations. The Free Press, New York
Porter M (1998) Clusters and competition: new agendas for companies, governments, institutions. In: On competition. Harvard Business School Press, Boston, pp 155–196
Porter M (2000) Location, competition, and economic development: local clusters in a global economy. Econ Dev Q 14(1):15–34
Porter M (2003) The economic performance of regions. Reg Stud 37(6&7):549–578
Porter M, Ketels C, Miller K, Bryden R (2004) Competitiveness in rural U.S. regions: learning and research agenda. Institute for Strategy and Competitiveness, Harvard Business School, Boston
Puppim de Oliveira JA (2008) Upgrading clusters and small enterprises: environmental, labour, innovation and social Issues. Ashgate, Farnham
Romer J (1986) Increasing returns and long-run growth. J Polit Econ 94(5):1002–1037
Schumpeter JA (1934) The theory of economic development. Harvard University Press, Cambridge, MA
Schumpeter JA (1950) Capitalism, socialism and democracy, 3rd edn. Harper & Row, New York
Sölvell Ö, Lindqvist G, Ketels C (2003) The cluster initiative greenbook. Ivory Tower AB, Stockholm
Swann GMP (2009) The economics of innovation: an introduction. Edward Elgar, Cheltenham
Tödtling F, Trippl M (2005) One size fits all? Towards a differentiated regional innovation policy approach. Res Policy 34(8):1203–1219

Part I
Food Security from the Asian Perspective

Chapter 2
Defining Japan's Food Security in East Asia

From the Perspectives of the Distribution Revolution, Environmental Degradation, and International Cooperation

Yonosuke Hara

Abstract This paper pointed out that the English phrase "food security" is routinely taken in Japan as dealing in advance with the contingencies of large-scale natural disasters and international conflict. The food stockpiling concept, in which Japan now leads the East Asian region, is precisely a food security scheme in the short-term sense for stabilizing supply from time to time. However, the phrase "food security" also has the meaning of a situation where the necessary food gets through to all the people on the planet. Food security in this sense can be said to be primarily a problem of the poor in developing countries, and a problem of their food purchasing power. It can be said that those in so-called absolute poverty in East Asia have certainly been decreasing in number, but by no means is the problem of poverty going away. It also clarified that such a poverty problem is profoundly linked to the degradation of natural resources, which can also be called the basis of agricultural production.

Therefore, it concluded that the food security problem itself in a long-term sense is emerging as the greatest challenge of today. Japan, which cannot survive without international society, must seriously consider what mechanisms it can use to combine the above two kinds of security, which differ in terms of their dimensions and time horizons. It is precisely in East Asia that Japan should demonstrate leadership in creating concepts regarding such mechanisms.

Keywords Distribution revolution • Environmental degradation • International cooperation • East Asia

---

Y. Hara (✉)
National Graduate Institute for Policy Studies (GRIPS), Tokyo, Japan
e-mail: yhara@grips.ac.jp

1 Interest in the Issues

Kindly allow me to begin this paper by all of a sudden talking of private matters. I travelled to Thailand for the first time in December 1973. This was a time almost exactly 1 month and a half after the "Student Revolution" which is described as a turning point in modern Thai history. On the trip, although we almost can't picture it now, in front of the "Thai Daimaru" department store, the sole piece of Japanese equity in downtown Bangkok, I witnessed students demonstrating for the boycott of Japanese goods. At the time, economic relations between Japan and Southeast Asian nations such as Thailand were greatly different to the present where agreement was able to be concluded on the ASEAN–Japan Comprehensive Economic Partnership.

From August 1975, I worked for a 2-year period as an expert in the agriculture division of the United Nations Economic and Social Commission for Asia and the Pacific (ESCAP) headquartered in Bangkok. Consequently, one of the tasks I was in charge of was to explore the possibilities for the realization of the Asian Rice Trade Fund concept. This was a period, immediately after the world food crisis at the beginning of the 1970s, when most Asian nations needed to secure a stable supply of the basic foodstuff, rice. Already around that time the results of the "Green Revolution" were beginning to be seen, it was when the designing and putting in place of regional rice trading schemes had become an issue. As part of the initiatives on this issue, with Thai friends I estimated what would be the optimum amount for rice stockpiles in the Asian region, and made consideration of a scheme to channel that stockpile and on what scale to countries with a rice deficit.

For East Asia as a whole, the production of rice via the introduction of high-yield varieties in the 1960s has expanded, and as a consequence for practically the whole region the real price has been following a downward trend. Therefore, in the same way as the era of Japan's old Agricultural Basic Act, the diversification of agriculture toward commodities for which a greater expansion of demand is anticipated has been progressing. At the same time as this, being pushed by the power of economic globalization, changes have been occurring which can be called a "revolution" in the structure of markets in which agricultural food products are being traded in every country. In this fashion, large structural changes have occurred over the last 30-plus years in the East Asian intraregional food economy. In spite of these changes, however, there is no doubt that the countries of Asia, starting with Japan, share the problem of advancing preparation for short-term fluctuations in food production. This is the very task I was assigned at ESCAP. In this aspect, the times have not changed at all. This unchanging issue can be called the problem of food security. Furthermore, it has begun to be acknowledged that global warming is the greatest challenge in the history of humankind in this century, and the pattern of monsoon rainfall in Southeast Asia also is changing greatly. In addition, salinization damage on the plains and landslides on slopes in mountainous areas have become evident in many places. Yet further, the deterioration of the agricultural productive resources of soil and water has been advancing strongly. Precisely the very issues of the long-term maintenance and preservation of food production capacity, which mostly had no attention paid to them 30 years ago, have become the fundamental problems for food security. In other words, the occasional short-term problems of a stable food supply which have continued to exist since time immemorial and the twenty-first-century long-term problems of the deterioration of agricultural resources have both become central issues of food security for East Asia and the entire world. And then the issue of Japan's food security also can no longer be considered one for Japan as one country alone. First acknowledging this fact dispassionately is a major premise for the debate.

2 The "Revolution" in Agricultural Produce and Food Distribution Organization and Systems in the East Asian Region

Owing to the high-level economic growth which was praised at one point as the "East Asian Miracle", in most countries in the East Asian region, particularly in urban areas a middle class with a high household income has emerged. In their consumer behavior for agricultural food products, they have changed the consumption style toward fresh produce, frozen food, and pursuit of safety and convenience, and eating out furthermore. Still more, technological innovation, such as the spread of temperature-control technology in the distribution of fresh foods has advanced. In many Asian countries, they have permitted the introduction of direct investment from overseas in the retail sector also. For this reason, the direct investment by private supermarket-related firms, including from Japan, has advanced greatly. In almost whole East Asia, in the systems and mechanisms for the distribution and trading of agricultural produce, great structural change in traditional configurations, from local traders through to wholesale markets, has occurred. This is certainly a great change, termed the "supermarket revolution" (Reardon and Timmer 2007). Supermarkets have today not only forayed into the major cities such as capitals, but also into the provincial cities of each country, and thus are coming to have a major impact on the food purchasing of rural people. The income of people in the provinces is still low when compared with the major cities, but even so there is no doubt that the income level of the majority of people is rising. The food sold in such supermarkets was initially vegetable oils and packaged foods, but has gradually come to include semi-perishables like dairy produce, including milk, and perishables like fruit and vegetables.

When the competition among supermarkets heightens, then what unavoidably comes to be necessary is the development of mechanisms to economize transaction costs for creating mechanisms for picking up a variety of foods and delivering them to the consumer. In other words, the procurement sections of each company have no choice but to demand new mechanisms to enable the picking up of more products of high quality. From the mechanism of storing stock at each retail location to distribute it, the establishment of centers which centralize the flow of goods to several retail locations will become essential. In this way, the traditional wholesale markets which have been the form of distribution in Asian countries have gradually decreased their importance, and in their place specialized centralized wholesale markets which the supermarkets have created are beginning to become central. Thus, as the economies of the countries of Asia go on integrating into broad networks of trade and direct investment, the accumulation and distribution channels of agricultural produce and food items also will continue transforming into broad-ranging ones connecting up national borders. That is, within agricultural produce and food items, the so-called intra-industry trade has begun to evolve.

The small farmers of Asian countries have also begun to get caught up in this change in distribution routes. For them, a method for continuously gathering goods towards supermarkets does not exist at the present, except the links to them in the form of contract farming. However, whether farmers can adequately respond to such great changes in distribution mechanisms is, as expected, a problem. Basically it may be said that, more than the operational scale of the area of farms under cultivation, the financial capital strength and education level which individual farmers possess are giving rise to differences in the responses of farmers. Naturally, to this day the majority of small farmers has depended on traditional distribution routes via bazaars and the like, but their importance in whichever country has begun to greatly decrease. This trend, amid the tide of economic globalization, can be called inevitable.

Let's take a little closer look at China. Entering the 1990s, the entry regulations for the distribution market and price controls were relaxed, and the monopolistic distribution systems by state organizations fundamentally collapsed. More specifically, a "partner trade" has been increasing where privatized wholesale firms circulate goods with wholesale firms in other regions. In addition, for the modernization of distribution, the central and local governments have begun constructing large-scale wholesale markets, such as centers for the distribution of goods, and processing centers. Thorough these processes, the modernization of the distribution system where the distribution and production of agricultural goods are integrated has progressed. In China also, undergoing the changes in the urban middle-class consumption style and the technological innovation within distribution, distribution firms which were rated as subordinate organizations for the distribution system to date, have made their appearance as important mainstays in distribution. Furthermore, organized retail "chain stores" have increased, which provide specialist services more freely, including supermarkets, convenience stores, drugstores and home centers, and business categories have highly diversified. In particular, from 1995 on, there have also been the moves of local governments for the active introduction of foreign capital, and the entries of such firms as Carrefour, Metro, Walmart, Daiei, Æon, and Ito-Yokado, have followed one after the other. In this way, within the retail sector the share of sales for state-owned and collectively owned enterprises, which was close to 80 % in the planned economy period, has fallen substantially (Shibata 2007). This is certainly a "supermarket revolution".

Moving on, let's review the history of changes in Japan's distribution systems for agricultural food produce. A major turning point for the agricultural produce trading and distribution systems in modern Japan, learning from the perishable agricultural produce distribution systems of other countries, was that the Wholesale Market Act came into effect in 1923. Through this development of legislation, wholesale markets were established around the country, then a mechanism was constructed to distribute goods from the producer to the consumer, with vegetables, fruit and marine produce basically going via these wholesale markets, and the antiquated "country markets" which had been in every part of the country until then were codified in law in stepwise manner (Suzuki 1990). Then after the war, with the enactment of the Agricultural Co-operative Act in 1947, single or multiple comprehensive agricultural cooperatives were organized at the city, town and village level all over the country. Additionally, the following were established: at the prefectural levels, business and economic federations and the like; and at the national level, the National Marketing Federation of Agricultural Cooperative Associations, the National Purchasing Federation of Agricultural Cooperative Associations, and the Central Union of Agricultural Co-operatives which can give guidance to all agricultural cooperatives. In this way was achieved a national mechanism for agricultural food produce to flow via agricultural cooperatives to wholesale markets.

The transformation of the trading and distribution systems which have been formed in this manner started to be visible from the beginning of the 1960s. At that time the distribution reform doctrine which was called the "wholesaler unnecessary doctrine" was actively being discussed. This distribution process short-circuit doctrine emphasized the importance of cutting out the intermediate stages linking the producer and the consumer. Also at that time business directly connecting to production areas to attempt to link producer and consumer without going through the agricultural cooperative organization began to appear. This period was one when talk was often heard that these direct business were hostile to agricultural cooperatives.

Regarding rice also, for which there had been strong government regulation since the wartime formulation of the Foodstuffs Control Act, a voluntarily marketed rice system was proposed, alongside the set aside policy of 10 % across the board nationally in 1969. This was the start of the so-called "great transformation to shake the world". Then the agricultural cooperative rice direct-sale business was started. More specifically, agricultural cooperative made the maximum use of agricultural cooperatives' rice-retailing certification, and gave the normal run of treatment for actually certified stores to those without the retail certification. Subsequently, it made active use of the rice-wholesaling certification which economic federations possess for retail-certified newcomers, expanded the agricultural cooperative-affiliated retail stores, and at the same time, regarding the "voluntarily marketed rice system", it eventually promoted cooperative society retail-certified newcomers. Moreover, in 1972, the Price Control Ordinance for rice was abolished, and essentially moved it over to free pricing. In 1978 the special cultivated rice system also appeared, which officially approved direct links between producer and consumer. In this way, Japan entered the period of rice direct from the production area officially becoming practicable. In other words, the portion which the nation mediated in large-scale wholesaling declined greatly, and in principle, it meant a move to a distribution liberalization era. Subsequently, via the 1996 Staple Food Act and the 2004 Revised Staple Food Act, the remnants of the Foodstuffs Control Act which was based on total volume control and official prices have been completely wiped away. In this way, the rice price formation system also moved over to one founded on market mechanisms. More specifically, the contract price at the Japan Rice Trading and Price Formation Center came to be the standard trading price at each stage of distribution from rice production to consumption.

Currently at any rate, regardless of whether for rice, vegetables, fruit, or marine produce, the trading and distribution channels have diversified. This diversified structure can be called "one-country plural systems" of traditional purchase and-salethrough agricultural cooperatives, and direct production – sale. Precisely on the point of the mechanism for trading and distribution of foodstuffs and agricultural produce, over the past decades Japan also has gone through virtually the same transformation as the countries of East Asia (Konno 2007; Tsujimura 2007).

3 Toward the Construction of a Food Security System in the East Asian Region

From the latter half of the 1970s when I stayed in Thailand down to today some 30-plus years later, the times have truly changed greatly. Talk of anti-Japanese political movements can mostly no longer be heard in Southeast Asia. As far as economic matters are concerned, Southeast Asia and Japan have deepened the integration in the aspects of trade and investment. Exactly the same thing can be said for food and other agricultural products. In other words, along with the progress in globalization and regionalization of market economies, the international structures concerning agriculture have also begun to exhibit changes which can be termed radical. With the information and telecommunications revolution, agricultural production by multinational firms has been evolving, and agribusiness has begun to globalize, transcending national boundaries. It is fair to say that this is a development-and- import formula where the food manufacturing capital creates production hubs abroad. Thorough rapid development of system via the import of processed foods where the importing side is involved in the production process, the closeness between East Asian agricultural production and consumption in Japan is furthering. Within the direct investment into Asia from Japan, the shares of the marine products and food manufacturing industries have risen rapidly. The proportion from Asia in the food import total value reached 37 % in 2005. Additionally, Japan's share in the food total export value of Asian countries has risen from 15 % in 1970 to more than 30 % today. Certainly this is for the reason that the Japanese stomach is linked to Asia. Multinational firms, also including Japanese firms, have concluded contracts with small farmers in the rural areas of the Asian region they have forayed into, and are producing raw agricultural produce and food agricultural produce. Outside of the format of such cultivation contracts, there is also capitalist management, namely plantations, where they lease large-scale areas of agricultural land and employ large numbers of workers. Whatever the case, in Asia, even more than traditional subsistence agriculture, commercial agriculture, which aims at sales to the global marketplace, has developed greatly.

Currently, Japan's food self-sufficiency rate has slipped below 40 % in calorie terms, and moreover is only at 28 % for the grain self-sufficiency rate. These are well-known facts. Such a trend in increase in imports is also a consequence of the development of commercial agriculture in the Asian region touched upon immediately before. What can be hoped for is that their supply to Japan over the long term with capital ties and contractual ties is certified as "approved national production", and the proposal of adding this to the total for the self-sufficiency rate has even begun to be talked about.

Here, I would like to point out an issue which is not much stressed regarding the problem of the shrinking of Japan's agriculture production. This is the impact which the trend of the strong yen after the Plaza Accord in the middle of the 1980s had in particular. There is not enough space to discuss the details, but in particular there should be no doubt that this strong yen was the largest macroeconomic factor which provoked the downsizing of land-using agriculture. Put simply, because the yen rate against the dollar was over twice as high for 20-plus years, in order for Japanese agricultural production to maintain the same international competitiveness as before the strong yen, raising productivity was necessary to lower production costs to less than half. For the small scale of management of Japanese farms, however, realizing such an improvement of productivity for a period of some 20 years was impossible. The consequence of that was the drop in the food self-sufficiency ratio. As in modern-day Japan, in an economy where the proportion which the agricultural sector occupies in total national income has become low, the economic activity of the non-agricultural sector, through the macroeconomic route of changes in the exchange rate, ends up having a one-sided influence on agriculture. This matter is not much emphasized in agricultural debate, but is a fact that should be acknowledged clearly. Then, actually, it is expected that the countries of East Asia will sooner or later also have no choice but to confront the same problem. The direct manifestation of that is the appreciation of the yuan which the United States and others have been forcefully requesting to China. It is fully expected that an appreciation of the yuan over too short a period will have a major impact on Chinese agriculture which has a similar agricultural structure to Japan.

As is well known, presently Japan is tackling the historical issue of agricultural policy reform, with an objective of concentrating support to the approximately 400,000 households practicing professional farm management, including 30–50,000 practicing corporate and community farming management. To that end, policy instruments have begun to be adopted for direct income payments to these "actors", taking as core the "Cross-Commodities Management Stabilization Measures". Explanation may not be necessary here, but this agricultural policy reform is something that aims at the securing of agricultural management entities that have international competitiveness, in response to the liberalization of agriculture obliged to within the WTO and a number of FTAs with regions or countries. Regarding such agricultural policy reforms, however, even if they were something that could prepare the necessary conditions for the resuscitation of Japanese agriculture, the fact that they are far from sufficient conditions must not be downplayed (for more details in this area please refer to the author's other work). Next, what I would like to emphasize here is the fact that such agricultural policy reform is not only the policy issues surrounding the border measures of the liberalization of trade in agricultural products, but also a response to structural change in a direction that penetrates the market principles of the domestic agricultural produce trading and distribution systems which have been sketched out.

Within the course of the expansion of reciprocal East Asian intraregional trade, the fact that trade in food and agricultural produce will continue expanding is already a trend which no one can deny. There has been China's WTO accession, and via the liberalization of trade, the trade in agricultural produce within the East Asian region has begun to push ahead in the rigorous process of competition and weeding-out of weaker agriculture within this region. Next, there is a forecast that, somewhat excessively, the profits of trade in agricultural produce within the East Asian region will be skewed toward China. This is because, in the intraregional trade in agricultural produce, the different phases of economic development, which is to say the disparities in labor cost, have stipulated the movement in the intraregional trade in agricultural produce (Suzuki 2007). On the other hand, in each country of East Asia, the problems of quality and safety of agricultural produce have unmistakably grown obvious. In these circumstances, once the broader market East Asian regional market is established, for agriculture more than the competition at the individual farm level, competition will continue to grow among communities and regions concerning brand agricultural produce. Nevertheless, the possibility will also be undeniable for this intraregional agricultural competition to evolve in a complementary and symbiotic direction.

In the contemporary world, in many regions, including Europe and the Americas, Free Trade Agreements (FTAs) have begun to mushroom. Whatever the region or country, food security is no longer a problem of one-country dimension. Yet for all that, regarding food and agricultural problems and resource and environmental problems, the argument for a full commitment to market principles which ignores the situations of individual countries is an extreme argument. Within the discussion surrounding the liberalization of trade in agricultural food products, a perspective balanced between consideration of this individuality of each country and fairness in the international dimension is required (Hara 2007). Next, even in trade liberalization regimes, there will continue to exist the inevitable issue of the instability of international markets. Not only trade liberalization, but also the concept of international commodity agreements and international food stockpiling which focuses on intraregional trade and market stabilization will be an important option. More specifically, the establishment of international food stockpiling organizations for spot commodities will come to be necessary (Kagatsume 2007).

In the East Asian region the problems are exactly the same. The policy debate needs to be expanded from a debate on food security within single countries to a common East Asian food security. Consequently, I would like to introduce the East Asian regional rice reserve and food security information systems which are now being discussed (Oba 2007). First there is the East Asia Rice Reserve System (EARR), and this secures rice reserves via the earmarking of spot commodity rice, and is a system for the reciprocal advancing of those rice reserves at times of emergency. One of the roots of the EARR was ASEAN food security cooperation based on the "Agreement on the ASEAN Food Security Reserve" (AFSR agreement) of October 1979. Next, there is the project of the ASEAN Food Security Information System (AFSIS). This had the objective of constructing a system for the effective exchange of information, gathering information on food agricultural produce supply and demand within the East Asian region, via the fostering of the human resources of the agricultural statistical information personnel of the member countries and the development of the information network system concerning the food supply and demand of the member countries. By proposing both of these, Japan has taken a major leadership role in their construction. In the WTO agricultural negotiations from 2001, Japan made a proposal arguing for the importance of Non-Trade Concerns, and incorporated therein a proposal for promotion of international cooperation on food security and international cooperation on food stockpiling in particular. This is the reason for attempting to realize these two concepts as the East Asian agricultural cooperation projects. More specifically, East Asian food security cooperation centered on EARR and AFSIS started at the end of 2001, essentially in a format in which Thailand's Ministry of Agriculture and Cooperatives with Japan's Ministry of Agriculture, Forestry and Fisheries and the Japan International Cooperation Agency lead the examination and implementation thereof.

It may be moving off the subject, but when I knew about the moves on EARR, I felt that the Asia Rice Trade Fund concept, with which I myself had been assigned at ESCAP, was finally going to begin in earnest. Moreover, regarding the moves on AFSIS, and while not the case that I myself was involved, I remembered that at the concerned party of ESCAP the starting-up of an intraregional agricultural information gathering project was being carried out.

The import of Japan's agricultural produce and food items is the importing indirectly the world's water resources. The concept of "virtual water trade", which focuses on this fact, has been set out. That is to say, the world's, and in particular East Asia's, water problems have begun taking on a life-or-death importance for Japan's food security. From such a perspective, the state and future of China's water resources in particular will come to be a major problem of vital significance for Japan also. At the present, the volume of water resources per capita in China is a mere quarter of the global average level, and for the northern part of China, which is a major grain-producing region, it is no more than one eighth. In China, from olden times "South-to-North Water Diversion Projects" to bring water to the north from the south have been implemented, and presently a similar project is underway. The greatest problem concerning the utilization of water resources in China, however, is the fact that the efficiency of utilization is crushingly low in comparison with Japan and the world. How seriously will China tackle this problem? (Shibata 2007). It should be no exaggeration to say that this very point is the most important challenge for the maintenance and preservation of the capacity and basis for food production of whole East Asia.

In Japan also there is the difficult problem of the maintenance and recovery of production capacity of abandoned arable land. Furthermore, it has been predicted that global warming will continue changing the Japanese archipelago from a temperate to a subtropical climate. Snow will continue disappearing as warming further progresses, and natural dams will continue disappearing. In the maintenance of food production capacity in Japan, the key will be coastal environmental and habitat restoration (Takemura 2007). Such opinions have also been put forward.

Japan's agricultural policy reform which has currently started to be undertaken takes as its basic direction the introduction of an international-standard "direct payment system". Income payments which target the narrowly-defined "actors" of large-scale operations, usually termed "industrial policy", are central to this agricultural reform. Direct payments, however, that target entire rural areas and which are termed "regional policy" and have at their core environmental payments, are also an enormously important policy. Direct payments are set with a target of maintenance and preservation activities for regional resources, including the farmland and water facilities fundamental to agricultural production. Next, direct payments target the agricultural management activities concerned with the preservation of the natural environment and habitat, including refraining from the use of pesticides and chemical fertilizers. Such environmental payments are supported by the recognition that environment and landscape cannot be bought with money. Won't the very enhancement of such broadly-defined environmental payments, not just mere measures for abandoned arable land, have to be taken as the basis for food security policy viewed from the long-term perspective?

The production base of a given country or region consists of the total capital stock including the natural resources which the ecosystem provides and the economic system which balances market economics and public intervention. Total capital stock except natural resources is constituted of artificial capital goods made by man and human capital enhanced by education. Both of these and natural resources will be greatly heterogeneous. Among the components of such total capital stock, I wonder if we can continue to expect optimistically that the human capitaland man-made capital goods, such as machinery, will be able to substitute for the shortage and degradation of natural resources. This very problem could be called the biggest problem upon which we must now reflect. The productive base of agricultural production is an ecosystem, the inner mechanism of which has not yet been clarified scientifically, and the changes and transformation thereof are irreversible changes which cannot be retrieved by human capability (Dasgupta 2007). In addition, for economic systems grounded only in market-fundamentalism, the maintenance and conservation of such an agricultural production base is impossible. To take such a simple fact lightly is no longer permissible.

4 Conclusion

From 2007 into 2008, amid the worldwide spread of lack of confidence in the US dollar, "speculative money", such as hedge funds, pension funds, and government funds, began to flow in large amounts from financial assets into oil and precious metals and then eventually into commodities such as grains. The expression "grain fund" is already becoming commonplace. Certainly, the price surge in such commodity markets is a "bubble that will eventually burst". I wonder whether it's a good thing to leave these flows of speculative money as they are, which are flowing into the basic commodities such as food. There is no doubt that this problem is on the agenda and must be discussed urgently. I remember that when I started studying economics, the "Commodities-Reserve Currency Proposal", to issue international money based upon the global stock of commodities including grains, was being suggested by international institutions. Today, at the beginning of the twenty-first century, such a bold concept will come to draw attention once again.

Against the backdrop of such intense moves, a war of a scramble for food is taking place. This scramble for food is beginning to become visible in three aspects. First, there is the scramble among nations. Second, there is the scramble between the two markets of energy and food. Lastly, there is the scramble for water and land between industry and agriculture. What is most important over the long term will be the scramble for what can also be called the basic resources for the food production. That is, the finiteness of food has undoubtedly begun to become visible. This is because the resources of water, soil and the global environment which are indispensable for food production have become more limited (Shibata 2007).

The English phrase "food security" is routinely taken in Japan as dealing in advance with the contingencies of large-scale natural disasters and international conflict. The food stockpiling concept, in which Japan now leads the East Asian region, is precisely a food security scheme in the short-term sense for stabilizing supply from time to time. However, the phrase "food security" also has the meaning of a situation where the necessary food gets through to all the people on the planet. Food security in this sense can be said to be primarily a problem of the poor of developing countries, and a problem of their food purchasing power. It can be said that those in so-called absolute poverty in East Asia have certainly been decreasing in number, but by no means is the problem of poverty going away. Next, it is also becoming evident that such a poverty problem is profoundly linked to the degradation of natural resources. The food security problem in this long-term sense is emerging as the greatest challenge of today. Japan, which cannot survive without international society, must seriously consider what mechanisms we can use to combine the above two kinds of security, which differ in terms of their dimensions and time horizons. We cannot target the whole world straight away, and it is precisely in East Asia that Japan should demonstrate leadership in creating concepts regarding such mechanisms.

There is also the matter of vociferous calls for "the hoped-for East Asian Community", but based on the differences in the stage of economic development and agriculture among the nations in the East Asian region, there first realistically needs to be a continued seeking of "gradual and incremental links". Amid this seeking, there must be the continuing realization of the construction of frameworks for international cooperative systems aiming at initiatives on short- and long-term food security problems.

References

Dasgupta P (2007) The economics of sustainability: human well-being and the natural environment. Iwanami Shoten, Tokyo (translation overseen by Ueta K) [in Japanese]

Hara Y (2007) The 'Farming' of the Northern Land and the Southern Islands: regional decentralization and agricultural policy reform. Shosekikobo Hayama Publishing Company, Tokyo [in Japanese]

Kagatsume M (2007) The merits and demerits of food trade liberalization and the significance of FTAs: can FTAs become the savior for food problems?. In: Noda K (ed) Living resource problems and the world: twenty-first century agriculture considered from the perspective of living resources no. 7. Kyoto University Press, Kyoto, pp 37–65 [in Japanese]

Konno S (2007) What are today's challenges for the direct marketing business of agricultural cooperatives? AT Journal No.10 [in Japanese]

Oba M (2007) Food security cooperation in east Asia. In: Shindo E, Toyoda T, Suzuki T (eds) An east Asian community pioneered by farming. Nihon Keizai Hyouronsha, Tokyo, pp 93–100 [in Japanese]

Reardon T, Timmer CP (2007) The supermarket revolution with Asian characteristics. In: Balisacan AM, Fuwa N (eds) Reasserting the rural development agenda, lessons learned and emerging challenges in Asia. Institute of Southeast Asian Studies, Singapore, pp 369–394

Shibata A (2007) The food scramble: the day when Japan's food gets left behind by the world. Nikkei, Tokyo [in Japanese]

Suzuki T (1990) The economics and institutionalism of the fruit and vegetable market. Gannando Shoten, Tokyo [in Japanese]

Suzuki N (2007) New perspectives on treatment of agriculture in the WTO and FTAs. J Rural Econ 79(2):49–64 [in Japanese]

Takemura K (2007) Lucky civilization: Japan will survive. PHP Institute, Kyoto [in Japanese]

Tsujimura H (2007) The challenges for the price formation and direct marketing of Kyoto Ayabe rice. AT Journal No.10 [in Japanese]

Chapter 3
Current Position and Future Direction of Agriculture in Northeast Asia

Shin-ichi Shogenji

Abstract This section discusses current situation and future perspective of food, agriculture and rural resources in Japan in the context of economic development in monsoon Asian countries. From food security concern importance of potential food supplying capacity rather than food self-sufficiency rates is emphasized. Regarding agriculture the challenge with top priority is to reconstruct sustainable paddy farming. Not only farmland size expansion for efficient operation but also vertical expansion of business size for enlarged job opportunities should be pursued. The vertical expansion includes diversification towards products marketing and food processing. As for rural resources such as irrigation facilities, the role of collective action for resource management is crucial in order to maintain the irreplaceable multi-functionality of agriculture. Finally policy implication based on the common characteristics of food, agriculture and rural resources among monsoon Asian countries is suggested, in particular, in relation to the ongoing WTO agreement

Keywords Food self-sufficiency • Sustainable paddy farming • Rural resource management • Multi-functionality of agriculture

1 Introduction

After the record-breaking price rise of cereals in 2007 and 2008 (Fig. 3.1), it is widely believed that the world food market has entered into a new stage in the following two aspects. Firstly, the market has become much more volatile than before. We also remember another acute price rise in 2010 and 2011. Secondly, the level of equilibrium price of basic foods has shifted upward, reflecting on-going changes in fundamental factors such as increasing food demand due to the economic growth in developing countries. Even after the end of rising phase, cereal price didn't return to the previous low level.

S.-i. Shogenji (✉)
Graduate School of Bioagricultural Sciences, Nagoya University, Nagoya, Japan
e-mail: ashogen@agri.nagoya-u.ac.jp

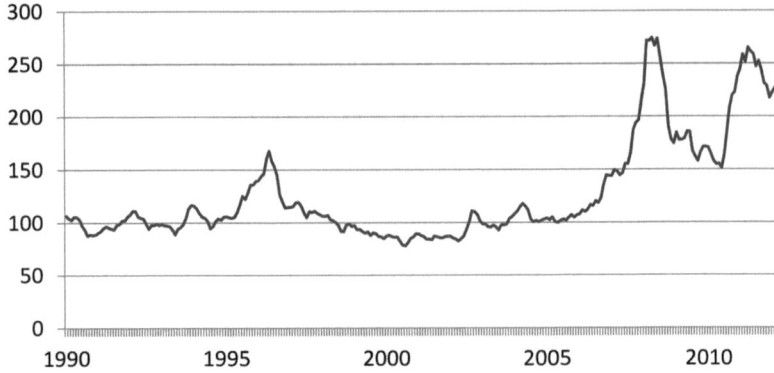

Fig. 3.1 Price index of world cereal market (Source: FAO, Food Price Index). Note: Monthly index. The average from 2002 to 2004 is set as 100

Table 3.1 Undernourishment in the world

	2007–09 (millions)
World	867
Developed regions	15
Developing regions	852
North Africa	4
Sub-Saharan Africa	216
Western Asia	18
Southern Asia	311
Caucasus and Central Asia	7
Eastern Asia	169
South-Eastern Asia	76
Latin America	43
Caribbean	7
Oceania	1

Source: FAO, The State of Food Insecurity in the World 2012

The changing world food market makes people across the world recognize the importance of food security. Needless to say, the price rise severely damaged poor households in developing countries (Table 3.1). This is the main reason why as many as 30 countries triggered their regulation on food export including embargo as its extreme case. It is quite natural that the government tried to protect its own nation. However it is also recognized that the regulation on food export accelerated price rise, especially in the case of rice.

Food security issue is important for people in food importing countries as well. In this regard we have to be cautious about the fact that regulations adopted by food exporting countries might damage food security in food importing countries. The changing world food market seriously poses food security problem to the government in food importing countries. And in this sense it has become quite important for northeast Asian region to understand its current position of food trade in the international context. Also it is important to design future direction of food supplying industry in a realistic manner.

In terms of position of food and agriculture, northeast Asian countries, that is Japan, Korea and Taiwan in this paper, share a couple of common characteristics. The first one is the position in food trade. Northeast Asian countries are now net importer of huge amount of food stuff, which is a consequence of continuous economic growth. It should be noted that developed countries in west Europe and those with European origin such as US and Australia are mostly net exporter of basic food, while developing countries by and large stay in a position of net importer. In this sense the food trade position of northeast Asian countries, that is, the position as already developed country with net import of food, is unique in the history of human beings.

The second commonness among northeast Asian countries is agricultural structure. Tiny farm size has long been major characteristic of farming in this region. And with rapid increase in per capita income, it became inevitable for farm business to expand its size in order to get comparable income. This proposition will sooner or later become applicable to other monsoon Asian countries which also share a similar structure of tiny size of farming, which in turn reflects historically high level of carrying capacity of paddy farming.

This paper focuses issues of food, agriculture and rural resources in Japan which has been front runner at least until recently in the post-war economic development. It will try to clearly show the characteristics of Japanese experience so far and then try to draw a few lessons for the future of northeast Asian region. The lessons might include negative ones which should not be followed after.

2 Declined Food Self-Sufficiency with Increased Food Consumption

Let us make a quick review of food and agriculture in Japan for the past half century. Firstly, Fig. 3.2 shows the rates of food self-sufficiency since 1960. Among three series the middle one, calorie base rate, is widely known in Japan. The rate had continuously declined and then became constant at around 40 % for the past two decades or so. This might give an impression that, firstly, Japanese agriculture gradually shrank for four decades and, secondly, then afterward has been keeping a certain level of production. Yet both are not correct.

As the indices in Table 3.2 shows, the aggregated agricultural production in Japan kept increasing until mid 1980s although products such as rice and potato decreased significantly. It was rather rapid change in people's dietary pattern that causes the decline in the rates of food self-sufficiency.

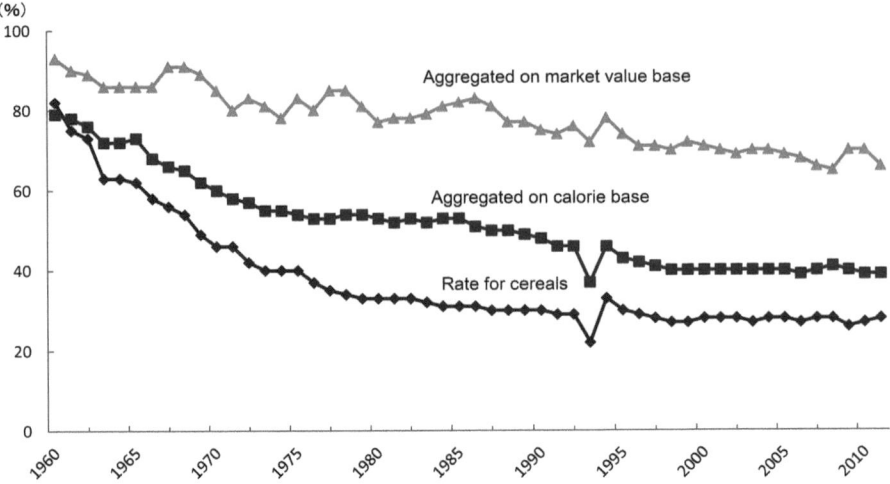

Fig. 3.2 Food self-sufficiency rates (Source: MAFF, Food Balance Sheet)

Table 3.2 Index of agricultural production

	Total	Rice	Wheat and barley	Pulses	Potatoes	Vegetables	Fruits	Livestock products
1960–64	100	100	100	100	100	100	100	100
1965–69	117	107	78	73	82	123	142	151
1970–74	120	94	27	64	60	135	184	205
1975–79	129	99	25	49	59	141	206	241
1980–84	129	84	44	49	63	145	199	280
1985–89	134	87	55	57	70	147	194	307
1990–94	128	81	38	40	63	137	172	313
1995–99	122	79	28	38	58	129	161	297
Apr–00	115	70	40	46	53	121	150	286

Source: MAFF, Production Index of Agriculture, Forestry and Fisheries

Table 3.3 shows how drastically the Japanese food consumption changed. In particular increase in livestock products is remarkable. These products were largely supplied by domestic production, but livestock production required huge amount of imported feed stuff such as corn. Also increased consumption of oil was supported by imported soybean.

Economic growth since mid-1950s changed people's food consumption drastically both in quality and quantity. In a word the dietary pattern in Japan was largely westernized. However, data in Table 3.3 also shows that, for some items such as dairy products, eggs and oil, per capita consumption already hit the peak and began decreasing. Perhaps, to some extent, the ageing demography now checks people's excessive intake. At the same time one can raise the following question. Does the recent change in dietary pattern in Japan suggest a possible peak level for people in northeast Asian region, or more widely for people of mongoloid race?

Table 3.3 Annual per capita consumption of food (kg)

	1955	1965	1975	1985	1990	1995	2000	2005	2010	2005/1995
Rice	110.7	111.7	88	74.6	70	67.8	64.6	61.4	59.5	0.55
Wheat	25.1	29	31.5	31.7	31.7	32.8	32.6	31.7	32.7	1.26
Potatoes	43.6	21.3	16	18.6	20.6	20.7	21.1	19.7	18.6	0.45
Starches	4.6	8.3	7.5	14.1	15.9	15.6	17.4	17.5	16.7	3.8
Pulses	9.4	9.5	9.4	9	9.2	8.8	9	9.3	8.4	0.99
Vegetables	82.3	108.2	109.4	110.8	108.4	105.8	102.4	96.3	88.1	1.17
Fruits	12.3	28.5	42.5	38.2	38.8	42.2	41.5	43.1	36.6	3.5
Meats	3.2	9.2	17.9	22.9	26	28.5	28.8	28.5	29.1	8.91
Eggs	3.7	11.3	13.7	14.5	16.1	17.2	17	16.6	16.5	4.49
Milk and dairy products	12.1	37.5	53.6	70.6	83.2	91.2	94.2	91.8	86.4	7.59
Fishes and shellfishes	26.3	28.1	34.9	35.3	37.5	39.3	37.2	34.6	29.4	1.32
Sugar	12.3	18.7	25.1	22	21.8	21.2	20.2	19.9	18.9	1.62
Fats and oils	2.7	6.3	10.9	14	14.2	14.6	15.1	14.6	13.5	5.41

Source: MAFF, Food Balance Sheet

Recently Japan has entered into the era of declining population. Therefore, with the declining per capita consumption, the denominator for calculation of self-sufficiency rate has been declining gradually. Therefore the almost constant level of self-sufficiency rate during the past two decades entails a declining tendency of domestic food production. This is confirmed through the data in Table 3.2. Constant rate of self-sufficiency does not mean stable situation of food production. Self-sufficiency rates must be interpreted carefully.

Facing declining agricultural production, the Japanese government decided to set target rates of self-sufficiency every 5 years in accordance with the Basic Law of Food, Agriculture and Rural Areas enacted in 1999. At present, target rates toward 2020 are set at 50 % for calorie base and at 70 % for market value base, respectively. The target rates are now under review towards the next Basic Plan of Food, Agriculture and Rural Areas which will be formulated by March 2015. At this stage it is difficult to tell the direction of ongoing review. Yet, towards March 2015, along with resetting process of self-sufficiency rates, so-called self-supplying capacity of domestic resource might become a point of argument.

There is no threshold level for the rate of food self-sufficiency under which the food security cannot be ensured. Also the following proposition is misleading. That is, the higher the rate, the more certainly the food security is ensured. For example, in 2009 the self sufficiency rate of cereals in India and Bangladesh were 104 % and 97 %, respectively. In the same year the rate was 26 % in Japan. But no one believes that the situation of food security in south Asia is far better than that in Japan. Simply the level of food self-sufficiency rate depends on people's dietary pattern.

Regarding food security, the more relevant indicator than self-sufficiency rate is absolute and potential supplying capacity of available resources to supply basic nutrition. The crucial question to be answered is whether this potential capacity, expressed in terms of per capita daily provision of calorie, will be sufficient or not for nation to survive any type of food shortage. Importance of absolute supplying capacity of domestic resources has been gradually shared among policy makers and even among some politicians.

3 How to Reconstruct Sustainable Paddy Farming

In Japanese agriculture there are subsectors, such as livestock farming and green house horticulture, which have been successful in farm size expansion and in getting income comparable to non-agricultural industries. In contrast a long-lasting problem remains in paddy farming. Japanese paddy farming could not overcome its farm size problem in spite of strong intention among policy makers. During the past half century the average size of rice growing farm became only doubled, while the average cow number of dairy farm, an example of successful intensive farming, grew up to more than 30 times larger (Table 3.4).

Average size of rice farming has been too small to get comparable income. As Fig. 3.3 indicates, per capita real income in 2005 was 7.7 times larger than that in

Table 3.4 Farm size expansion

		1960	1970	1980	1990	2000	2010
Rice growing farm size (a)		55.3	62.2	60.2	71.8	84.2	105.1
Dairy cows and calves (heads)		2	5.9	18.1	32.5	52.5	67.8
Farmland size (ha)	Other regions	0.77	0.81	0.82	1.1	1.21	1.42
	Hokkaido	3.54	5.36	8.1	10.8	14.3	21.5

Source: MAFF, Agriculture Census

Fig. 3.3 Per capita real GDP and population (Source: Cabinet Office, Annual Report on National Income and Ministry of Internal Affairs and Communications, Population Census). Note: Real GDP at 1990 fixed price is shown

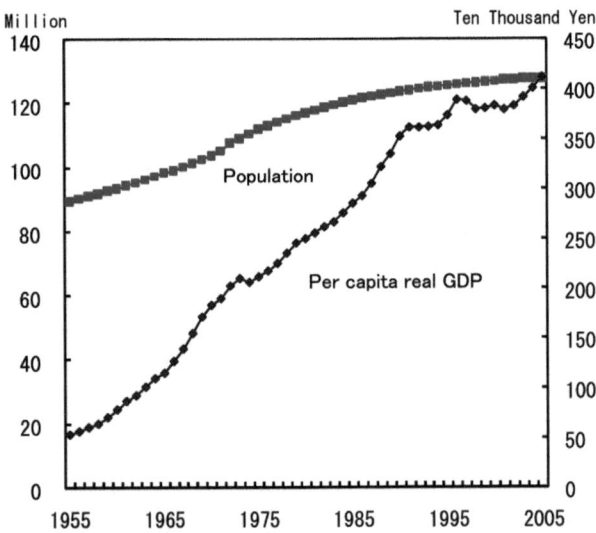

1955, the memorial year when the so-called high economic growth era had started in Japan. The majority of paddy farm households got their income mainly from off-farm job opportunities. They chose to continue their lives in rural areas as part-time farmer. This farmers' adjustment to economic growth was quite rational, taking into account widely spread job opportunities even in rural areas. Also modern equipments such as transplanting machine supported part-time farmers' holiday operation.

The part-time farming was not only rational but also stable until recently. However, this type of farming has been rapidly losing its sustainability mainly because of lack of successors. Along with aged farmers' retirement the amount of farmland to let has been increasing. This means that a good opportunity for farm size expansion is now provided. Whether this change of situation will work in favor of farm size expansion or not depends on a number of factors.

Among others the key factor must be consistent and stable policy which supports core farms such as full-time family farm, hamlet-based group farm and corporate type of farm. These core farms would expand their farm land size, if it is judged by them as profitable to do so. Regrettably actual process of agricultural policy in recent years was not consistent or stable enough. Rather there has been a sort of swinging situation where policy set affecting farm income changes frequently. One might criticize that agricultural policy in Japan became most serious risk factor of farming.

The main reason for the swinging situation is political instability. Not only actual change in ruling party but also election campaign for getting power tends to create unstable situation of agricultural policy. At the same time, basic and overall system for policy making should be carefully examined. There is the Basic Law for agricultural policy making in Japan. Current law was enacted in 1999. Also the government draws up the Basic Plan every 5 years in order to materialize the ideas of Basic Law as concrete policy measures. However, in spite of this formal procedure, additional new policy measures are occasionally created in the form of administrative guidance and/or budget item. This system brings about frequent change in policy measures. There have been cases that newly introduced policy measure by, for example, new budget item is inconsistent with provision of existing law.

One can call this system as loose system. When political situation is stable, there might not be serious inconsistency between existing lawful structure and policy set created through loose system. Rather flexibility and quickness given by this loose system might be appreciated and even enhance the efficiency of agricultural policy as a whole. Yet, it should be noted that loose system under politically unstable situation may create unpredictable swinging policy measures inconsistent with existing ones.

Let us think about another necessary condition for stable agricultural policy. That is consistent support by people as consumers and tax payers.

It is well known among professionals that farm size expansion will reduce production cost significantly. Figure 3.4 shows relationship between size of planted area and average production cost. Policy resource mobilized for farm size expansion will surely lead to a lower cost which in turn will result in lower food price and/or

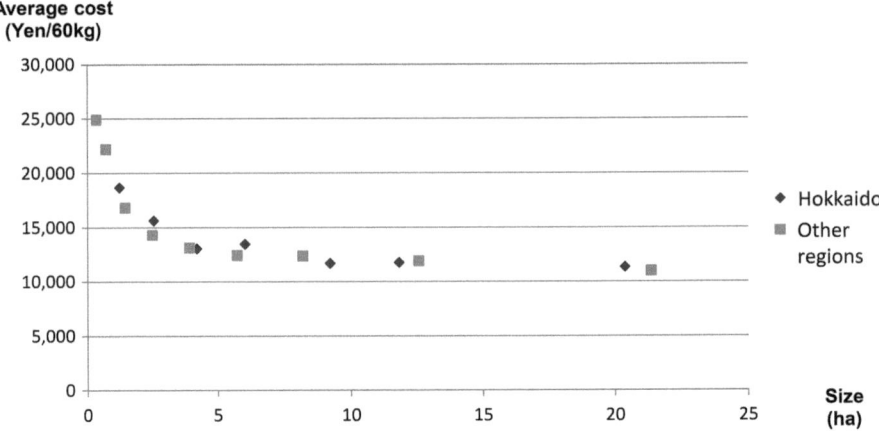

Fig. 3.4 Size and cost of rice production (2011) (Source: MAFF, Production Cost of Rice and Wheat)

decreased policy support expenditure. These policy effects would benefit people as consumers and/or people as tax payers. It is important to convey this sort of information to people outside farming community. For long-lasting and powerful agricultural policy nation-wide correct understanding about this relationship is essential.

For viable paddy farming as an income source, farm size expansion is undoubtedly basic condition. But in order to strengthen its sustainability through getting more value added and attracting young successors, vertical expansion of business size is also effective. The vertical expansion includes introduction of intensive farming items such as vegetables, high quality fruits and mushroom. This can be interpreted as a more intensive utilization of expanded farm land.

Also farmers might extend their business activities, introducing some elements of food industries. Food industries here mean food manufacturing, food distribution and restaurant industry. Nowadays it is not rare for farmers to process their products and/or to use their product in farmer's restaurant. Also it is fairly common that farmers directly sell their farm products to their customers, mainly using information technology. In terms of industry categorization these farms which sell products for themselves engage in food distributing business.

In the previous chapter change in dietary pattern was discussed in terms of composition of food stuff category. It is appropriate now to point another aspect of change in dietary pattern. This is increased use of processed food and increased frequency of eating out. Currently less than 20 % of consumers' expenditure on foods and drinks goes to unprocessed products such as vegetables, meats and rice. Over 50 % was spent to purchase processed food and around 30 % spent for eating out. Accordingly a large amount of value in food is now created in food industries which function in the downstream of agriculture and fisheries as food stuff industry.

Table 3.5 Working population of food-related industries

		1970	1980	1990	2000	2010
Number (ten thousand)	Agriculture and fisheries	987	596	430	320	309
	Food industry	509	643	723	804	792
	Processor	106	115	138	143	119
	Distributor	244	299	333	382	345
	Restaurants	159	229	253	280	328
	Total	1496	1239	1153	1124	1103
Percentage (%)	Agriculture and fisheries	66	48.1	37.3	28.5	28
	Food industry	34	51.9	62.7	71.5	71.8
	Processor	7.1	9.3	12	12.7	10.8
	Distributor	16.3	24.1	28.9	34	31.3
	Restaurants	10.6	18.5	21.9	24.9	29.7
	Total	100	100	100	100	100
Total working population		5259	5581	6168	6298	5961

Source: Ministry of Internal Affairs and Communications, Population Census

Growing presence of food industries can also be confirmed by changing composition of working population among industries shown in Table 3.5. Forty years ago ten million people worked in agriculture and fisheries while five million worked in food industries. Afterward employment in food industries grew significantly and nowadays eight million people work in food industries while three million in agriculture and fisheries. It is reasonable strategy for farmers to extend their business wing to get much more value added in downstream industries.

Among food industries, processing business is densely located in local cities and rural areas. As well as agriculture, food processing business is not highly profitable on average. But it is fairly stable against up and down of economic condition. Stable job opportunities will be one of the essential conditions for a matured society in which one cannot optimistically expect a high rate of economic growth experienced in the past.

In near future agriculture in connection with food industries, in particular, food processing industry in rural areas will be able to contribute as suppliers of stable job opportunities. If this situation spreads widely in rural Japan, one can say as follows.

In the long history of social and economic development since Meiji Restoration, rural community has been contributing in a variety of way. But among others contribution through providing massive workforce to secondary and tertiary industries was enormous indeed. However, now, the Japanese society stands at the turning point. Today rural community in Japan is expected to contribute to the stability of society in the opposite way, that is, by providing stable job opportunities, although not quite massive perhaps, for people even from urban areas.

4 Rural Resources for Future Generations

Farming with expanded land size will require some change in rural system of resource management. Regarding modernized Asian agriculture, in particular, paddy farming, its basic nature might be depicted as a two-tier structure (Fig. 3.5). Namely, the upper tier is business one which constantly interacts with market economy and the lower tier is communal one with collective action for maintenance of local common resources. As farm business on upper tier grows larger with modern machinery equipment, adaptive alteration becomes necessary for the system on the lower tier.

The function of lower tier normally takes a form of collective action by farmers. An example is collective action for maintenance of irrigation facilities such as canal and reservoir. It is quite common in paddy farming areas that farmers in a hamlet meet together one morning just before transplanting season in order to dredge canals and make repair if necessary. Usually one male adult is called from each household. Collective actions in a hamlet, and sometimes in a wider village area, cover many aspects of rural life. For instance, farm road maintenance, shrine maintenance, fire brigade, and social event such as festival and funeral are operated based on communal collective action.

Rural resources maintained by local people can be characterized as local commons under communal control. Different from the commons described in "The Tragedy of the Commons" by Hardin (1968), which broke down due to a rational and selfish behavior of members, local commons in rural Japan, and in monsoon Asian rural areas, has been kept for a long time up until present. The key to this continuity is communal rules of control which enable local commons to be transferred over generations.

However, composition of community member in rural area has changed significantly. In old days, shortly after the land reform which was completed in early 1950s, farm size and crops produced are similar among hamlet members. Hence, it was quite understandable for members to participate in communal activities. If the hamlet consists of 30 farm households, then, by participating in some collective action, each household makes 1/30 contribution to and gets 1/30 benefit from it. This relationship makes communal rules easily acceptable for members.

Today membership of hamlet is highly diversified especially in terms of involvement in farming. Usually majority of farm households get income from off-farm job. Also, there are ex-farmers who live in a hamlet as a land owner. In some cases land owner does not live in the area. Meanwhile a few household run their expanded farm as full-time farmers. However, even among full-time farmers main product for income earning might be different from each other.

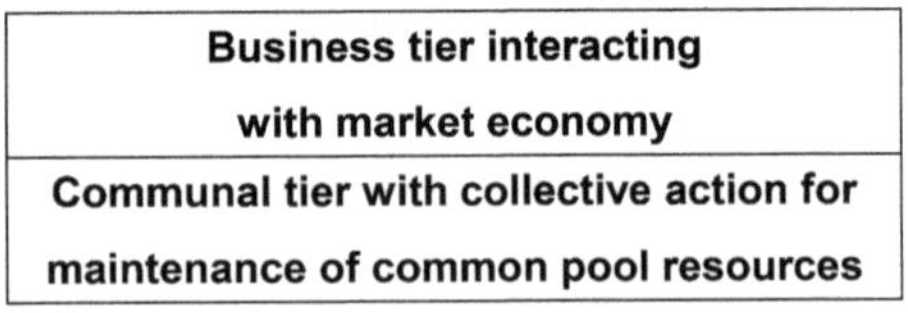

Fig. 3.5 Two tier structure of Japanese agriculture

Business tier interacting with market economy
Communal tier with collective action for maintenance of common pool resources

In short, rural community has changed from traditional homogeneous one to highly heterogeneous one. Under this heterogeneous structure, it has been becoming difficult for rural community to force its members to participate in communal activities as an a priori duty. Among members with varying involvement in paddy farming the balance between contribution and benefit is no longer self-evident to the members.

However, it is necessary and even rational for rural community in even today's Japan to organize collective action to maintain rural resources in good condition. Falling into inferior Nash equilibrium or prisoners' dilemma, the commons described by Hardin collapsed. But regarding actual commons in rural areas, win-win relationship can be achievable and should be obtained with coordinated equilibrium through cooperative communication among members.

It has been a historically given rules that controlled local commons. But in the future it will be an explicit consensus that maintains rural resources in a good condition. Sometimes patient communication will be necessary. It might be effective for consensus building to make an adjustment of burden on members among different communal activities. Also persuasion for avoiding myopic personal comparison between cost and benefit will be necessary for sustainable resource management.

Finally let us look at rural resource issues from a broader perspective. It is well known that the concept of multi-functionality of agriculture has been widely accepted both in EU countries and north-east Asian countries. Although type of farming is quite different between two regions, agriculture provides society with a variety of valuable byproducts in a form of external economy. The reason for this commonness lies in a similar structure of space use in rural areas. This can be called as multi-purpose use of rural space. This similarity in space use implies that large population in both EU and northeast Asian countries can actually enjoy multiple functions of agriculture on site.

As Fig. 3.6 shows in a simplified way, rural space in Japan, as well as in other northeast Asian countries, is used in three-dimensional manner, namely, as a space for primary industry, a space for residential place of households including non-farm households, and a space for visitors from outside community. This structure seems to be natural for people in EU and north-east Asian countries. It can be applied by and large for people in aged countries with lengthy history of social development. Yet, the situation is quite different in newly developed countries such as US, Canada, and Australia as shown in Fig. 3.7.

In aged counties almost all the available spaces were developed in early stage of their history. Therefore, society was forced to make use of given space more intensively for different purposes at the same time. In Japan majority of residents in rural areas are non-farm families, some of which have moved to the village from urban area. Also people in urban areas frequently visit rural areas as holiday makers or homecoming relatives. These dwellers and visitors can enjoy multi-functionality of agriculture such as landscape, natural creature, and traditional culture on site. Indeed, the concept of multi-functionality can be socially meaningful only with its repeating users.

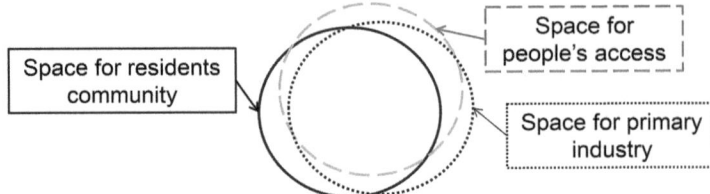

Fig. 3.6 Multi-purpose use of rural space in Japan and Europe

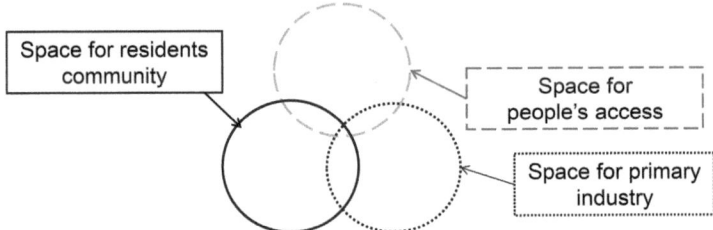

Fig. 3.7 Abundant space for each purpose in US and Australia

Japanese society has many things to be changed. Of course things in agriculture and rural areas are not exception. But at the same time Japanese society has many things not to be changed and to be handed over to the future generations. Among others this chapter stressed the importance of collective action for rural resource management on the one hand and the irreplaceable value of byproducts of agriculture based on multi-dimensional use of rural space on the other.

5 Concluding Remarks

Taking into account common elements in agriculture in northeast Asian countries, more intimate academic interchange of information and ideas will be highly beneficial for each country. This paper summarized the Japanese situation. It is author's pleasure, if it gives readers a small hint for considering his or her own nation's agriculture and agricultural policy.

Let me add one more thing. Among common elements, that of trading position of basic food strongly suggests the necessity to express our possible common interest in policy-making in an international context. In relation to this suggestion, let us remind ourselves that, after the conclusion of GATT Uruguay Round negotiation, the WTO member countries have been adjusting their agricultural policy in accordance with the WTO agreement. In particular policy measures which might lead to increase in agricultural production have been rather restrained.

Of course it is member country's obligation to observe the existing rules. But, at the same time, the rules are rules and can be, and should be modified, if appropriate. In this regard a sentence in introduction of this paper is worth repeating.

The food trade position of northeast Asian countries, that is, the position as already developed country with net import of food, is unique in the history of human beings.

This paper was written based on the author's keynote speech in 2014 Taiwan-Korea-Japan International Conference on Agricultural and Resources Economics held on August 13, 2014 at National Taiwan University in Taipei.

References

Food and Agriculture Organization of the United Nations (FAO) (2012) The state of food insecurity in the world 2012. FAO, Rome

Hardin G (1968) The tragedy of the commons. Science 162:1243–1248

Chapter 4
Analysis of China's Food Supply and Demand Balance and Food Security

Yongfu Chen and Fengying Nie

Abstract China, as a developing nation with a population in excess of 1.3 billion, supports more than 20 % of the world's population on less than 10 % of the world's cultivated land, and if a problem arises for China's food security it will become a profound issue for the global food market which cannot be ignored. In recent years China's total production volume for food and production volume per unit area have been on an upward trend, but food supply and demand is still going through a somewhat tight situation. As the supply volume shortfall is large for some food crops, the possibility is high of it having a great impact on China's food supply.

From such a background, in this paper we analyzed the changes in China's food production, consumption and trade and the causes impacting thereon, and also undertook examination of the forecasts for the food supply and demand situation. Building on that, we indicated the following three points as issues for China's future food security, based on the analytical results.

First, regarding the food self-sufficiency ratio, it is important to set the food self-sufficiency ratio at a realistic 80–95 % as a concrete policy objective, making appropriate adjustments after assessing the situation for the international food supply and demand balance. Second, via measures including the development and dissemination of agricultural technology, the putting in place of agricultural infrastructure, and the prevention of disease and insect damage, it is necessary to raise the total utilization efficiency rate of water and land resources and constantly improve the production volume per unit area. Third, at the same time as developing the food futures market in focused fashion, it is necessary to establish price changes via market mechanisms. Moreover, in addition to taking a series of countermeasures domestically, the implementing of a multifactorial import strategy for agricultural produce is also important.

Y. Chen (✉)
College of Economics and Management, China Agricultural University, Beijing, China
e-mail: chenyf@cau.edu.cn

F. Nie
International Information Division, Agricultural Information Institute,
Chinese Academy of Agricultural Sciences, Beijing, China

Keywords Food production capacity • Food self-sufficiency ratio • Food supply-and-demand forecast • Food security

1 The Background to the Research and the Setting of Agendas

China, as a developing country with a population of 1.3 billion, supports more than 20 % of the world's population on less than 10 % of the world's cultivated land, and has been contributing to global food security. If a problem arises for China's food security, it can have a profound impact on the global economy (food importing nations in particular). China's food security problem is consistently garnering attention, and is a factor which prompts argument.

The argument surrounding China's food security has been continuing for many years, and can be divided into the optimists' camp, the pessimists' camp, and those in the middle. Scholars within China generally belong to the optimistic and center camps, and there are many overseas scholars in the pessimistic camp (for example, Brown 1995). There is a strong connection among such arguments as that China's per capita resources are low and the population is high, and the shift in the structure of food consumption accompanying its rapid economic growth.

The total production for food in recent years and the yield per unit area have been on an upward trend, but China's food supply and demand is still going through a somewhat tight situation. In addition, the frequency of the occurrence of natural disasters in China is on an upward trend, the water resource crisis is deepening further, and disasters such as drought and flooding are increasing in certain areas with global warming. As the supply shortfall is large for some food crops, it is having a great impact on China's food supply. If a problem occurs at the distribution stage for these crops, then the future security of the food supply will become unsettled. Suffering the impact of bio-energy and a surge in energy prices, and shortages in food stocks on a global scale, international food prices in the most recent few years have also risen substantially, and China's food prices have sharply increased. Furthermore, with there also being the impact of extreme weather, China's food supply problem is being made prominent, and is becoming a common matter of concern both inside and outside the country. Therefore, in this section we would like to examine primarily the impact which this and China's food supply and demand balance have on food security. Moreover, as a supplementary explanation, we will add the following two points: (1) wheat, rice, maize, soybeans and miscellaneous grains are included in what are termed "foodstuffs" in China; and (2) for what are termed "grains" in China, we denote the "foodstuffs" other than soybeans. We will use these definitions in this section.

Below, in addition to analyzing the changes in China's food production, consumption and trade, and the causes thereof, we will examine food security strategy, after forecasting the food supply and demand situation.

2 Analysis of the Changes in Food Production, Consumption and Trade and the Factors Having an Impact

2.1 The Trends in Food Production Volume and the Changes in Food Consumption

The area of China's cultivated land at the beginning of the twentieth century had already reached 90.43 million hectares (henceforth abbreviated to "ha"), the total production volume was 193 million tonnes, and the yield per unit area was 2.134 tonnes/ha. In 1950 China's area of cultivated land for food expanded to 114.41 million ha, but the total production volume fell to 132.13 million tonnes, and the production volume per unit area decreased to 1.155 tonnes/ha (Xu 1981). The area cultivated for food in 2006 decreased somewhat to 105.49 million ha, and was used for the construction of roads and housing accompanying urbanization. Meanwhile, the total food production volume expanded to 497 million tonnes, and the production volume per unit area rose to 4.716 tonnes/ha. Which is to say, China's food production volume increased 2.57-fold compared to the beginning of the twentieth century, and it may be said that the increase in production volume was due to the rise in the yield per unit area.

As can be seen from Table 4.1, marked changes can be seen in the structure of per capita food consumption in China from the 1950s to 2006. The per capita volume of food-use foodstuff consumption has changed from an increasing to a decreasing trend, and the per capita volume of vegetable consumption has also been decreasing. Meanwhile, the consumption volumes of food-use vegetable oil, pork, beef and mutton, poultry, eggs and their processed products, aquatic products, milk and its processed products, and liquor substantially increased. In particular, the per capita consumption volumes of beef and mutton and milk and its processed products have increased greatly. The changes in the structure of this per capita food consumption inevitably had the result of bringing about an increase in the food volumes necessary for the increase in demand for the volume of food for animal feed and per person.

2.2 Analysis of the Changes in Trade in Food and Food-Use Oil

Regarding the net volume of grain exports for China the extent of change has been an unremitting increase. In particular, from the 1990s, for certain years, the extent of change has climbed to close to 20 million tonnes. Looking at the breakdown by time period, from 1867 to 1950 China was a net-importer of grain, but switched over to a net-exporter of grain in the period 1950–1960. Then, for the period 1961–1984 it once again became a net-importer, but from 1984 on, it has seesawed between being a net-exporter and net-importer of grain.

Table 4.1 The change in the structure of food consumption in China (Unit: kg/person)

Year	1952	1957	1962	1970	1978	1980	1981	1985	1990	1995	2000	2005	2006
Food (unprocessed products)	197.7	203.1	164.6	187.2	195.5	213.8	219.2	254.4	242.2	221.9	203.5	169.6	163.6
Vegetables	–	–	–	–	–	–	–	–	135.2	108.1	109.7	107.6	108.5
Food-use vegetable oil	2.1	2.4	1.1	1.6	1.6	2.3	2.9	5.1	4.3	5.1	6.3	6.3	6.7
Pork	5.9	5.1	2.2	6	7.7	11.2	11.1	14	12.6	12.5	14.6	16.1	17.5
Beef and mutton	0.9	1.1	0.8	0.8	0.8	0.8	0.8	1.3	1.5	1.2	1.8	2.3	2.5
Poultry	0.4	0.5	0.4	0.3	0.4	0.8	0.8	1.6	1.8	2.5	3.6	4.4	5.9
Eggs	1	1.3	0.8	1.3	2	2.3	2.4	5	3.7	5.1	7	7.5	7.4
Aquatic products	2.7	4.3	3	2.9	3.5	3.4	3.6	4.9	3.6	5.1	6.2	7.9	8.3
Milk and processed products	–	–	–	–	–	–	–	–	2	1.8	3.5	5.9	9.6
Liquor	1.1	1.4	1.1	1.3	2.6	3.4	4.4	7.7	7	7.5	8	9.8	9.5

Source: Compiled from the *China Statistical Yearbook*, 1981, 1985, 1996 and 2007 editions

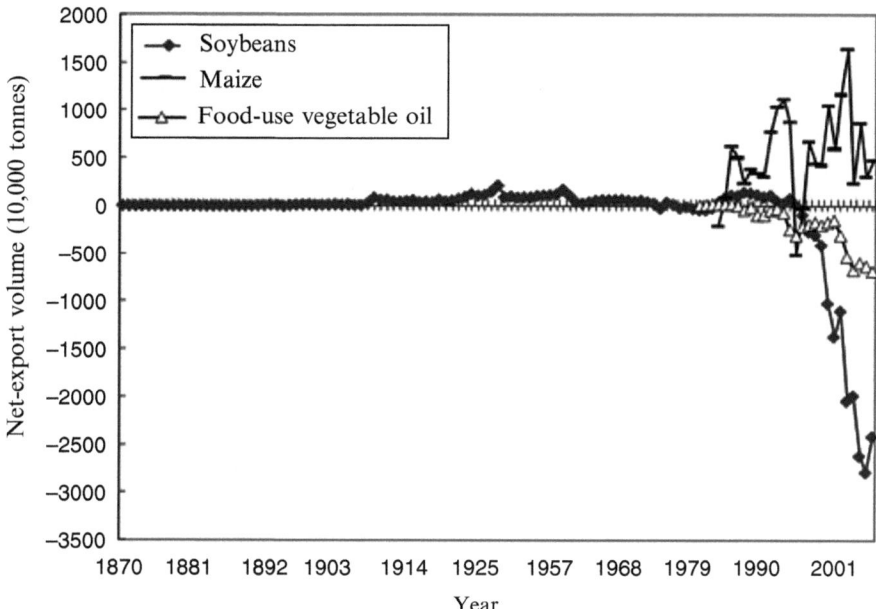

Fig. 4.1 The trends in trade volume for soybeans, maize, and food-use vegetable oil (Source: As for Table 4.1)

Furthermore, in the years that China was a net-importer, the net volume of imports was quite large. In particular, from the middle of the 1990s, owing to the rapid change in China's net-import volume for grain, countries all over the world, and especially the net-food-importing nations of East Asia, came to harbor concerns about China's food security problems.

As can be seen from Fig. 4.1, from 1984 on (with the exception of 1995 and 1996), China has continuously been a net-exporter of maize, and in 2003 the net volume of maize exports reached 16.39 million tonnes. As can be seen after analyzing the historical changes for food production, the upshot of the growing area for maize and the yield per unit area continuing to expand without let-up was that China became a net-exporter of maize.

Before the mid-1990s (with the exception of the 1970s), China had for a long time been a net-exporter of soybeans. After China became a net-importer of soybeans in 1996, the net-volume of soybeans imported continued to increase rapidly, and reached 27.88 million tonnes in 2006. Through the net-import volume of soybeans continually increasing, the concern of countries all around the world about China's food security has deepened yet further.

Suffering the effects of the limitations in natural resources, the increase in the per capita volume of vegetable oil consumption, and the rapid increase in population, China became a net-importer of vegetable oil in 1986. Subsequently, in 2006 China's net-import volume of vegetable oil increased to 6.31 million tonnes. For the period from January to October 2007, the net-import volume of food-use vegetable oil in China reached 6.84 million tonnes, and within that the net-import volume of palm oil was the largest, followed by that for soybean oil.

2.3 Analysis of the Limiting Factors Affecting the Food Supply

For a long time, among the limiting factors for the food supply in China, with the exception of price factors, the most fundamental factors were per capita natural resources and natural disasters. As shown in Table 4.2, dramatic changes in China's population were seen in every period. Taking the year of 1661 as the starting point, the population increased from 14,030,000 to 1,314,480,000 in 2006, but the increase in the area cultivated has not kept pace at all compared to the population increase. As a result China's per capita cultivated area has greatly decreased, from 27.63 mu

Table 4.2 The changes in China's total population and area cultivated

Period	Year (western calendar)	Per capita cultivated area (mu per person)	Total population (10,000 persons)	Cultivated area (10,000 mu)
Western Han	2	13.88	5,959	82,705
Eastern Han	105	13.74	5,326	73,202
Sui	609	121.37	4,602	558,540
Tang	726	34.78	4,142	144,039
Northern Song	1021	26.33	1,993	52,476
Ming	1381	6.13	5,987	36,677
Qing	1661	27.63	1,403	38,777
	1724	34.11	2,611	89,065
	1753	6.89	10,275	70,811
	1766	3.56	20,810	74,145
	1812	2.19	36,169	79,153
	1833	1.7	39,894	67,999
	1863	1.86	40,495	75,176
	1872	2.29	32,956	75,554
	1887	2.49	33,759	84,084
	1900	2.31	36,681	84,778
	1910	3.95	36,815	145,524
Republic of China	1916	3.12	40,950	127,689
	1934	2.66	46,215	122,837
	1947	3.05	46,280	141,073
People's Republic of China	1950	2.73	55,196	150,534
	2006	1.39	131,448	182,700

Source: Compiled from Yu and Teng (2000), and Xu (1981)

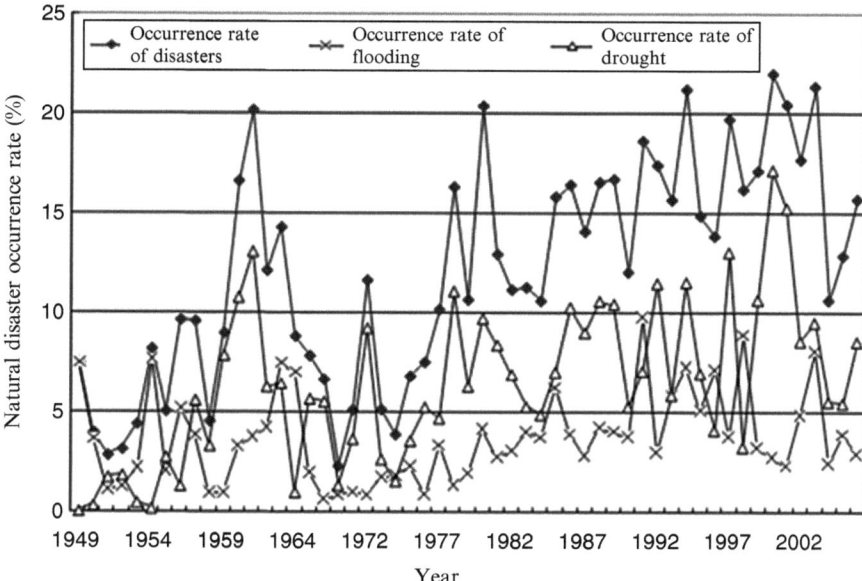

Fig. 4.2 The historical trends in the frequency of occurrence of natural disasters in China. Note: Natural disaster occurrence rate = disaster area/ total crop-planted area × 100 % (Source: Compiled from the online natural disaster database of the Department of Crop Farming Administration of the Ministry of Agriculture of the People's Republic of China)

in 1661 to 1.39 mu in 2006 (1 ha = 15 mu). In order for China to go on supporting 1.3 billion by itself, it has no option other than continually to raise the food production volume per unit area. That is, they have to plan the enhancement of the overall capacity for food production, including the upgrading of agricultural infrastructure and the raising of agricultural production.

As can be seen from Fig. 4.2, a continually rising trend is apparent for the occurrence rate of natural disasters in China. Among the various kinds of natural disaster, the occurrence rate for drought is the highest. Marrying the natural disaster occurrence rate up with the years with a net-import volume for grain, when the natural disaster occurrence rate rose to 21.2 % in 1994, the net-import volume for grain for 1995 rose to 19.75 million tonnes. Moreover, as the natural disaster occurrence rate for 2003 was 21.3 %, for the net-import volume for grain there was a net-export of 19.91 million tonnes in 2003, but this switched over to a net-import of 4.96 million tonnes in 2004. For China in the future there is the possibility of facing frequent extreme weather and natural disasters accompanying global warming. If serious natural disasters occur, the probability is high that China's trade in grain will fluctuate greatly.

3 Forecasts for Food Supply and Demand

3.1 Past Results of Forecasts Relating to China's Food Supply and Demand

Regarding China's food supply and demand, Lester R. Brown forecast that the net-import volume for food would be 151 million tonnes in 2010, 258 million tonnes in 2020, and 370 million tonnes in 2030. The food net-import volume for 2010 which Brown forecast is close to the results of the forecast of Japan's Overseas Economic Cooperation Fund (OECF). The food net-import volume for 2020 which Brown forecast, however, differs by roughly 200 million tonnes compared to the results of other studies. Furthermore, in contrast to the 370 million tonnes for the food net-import volume for 2030 which Brown forecast, the volume forecast by the Economic Research Institute of China's State Development Planning Commission was a mere 63 million tonnes.

Regarding the cause for the results of overseas studies and forecasts differing, there were prior studies, including Fan and Sombilla (1997), Qu (1997), and Barney et al. (1999). According to these studies, the cause was the difference resulting from the forecast volume of production. That is to say, the food supply forecast was extremely fraught.

Other than that, the hypotheses, model parameters, model structures, and the linkage relationships between agriculture and other industries used in each of the forecast models differed, and the differences became the cause of the forecast results disagreeing. Within the forecast work which was specifically undertaken, namely within the specific simulated policy proposals, such matters as the supply price response, the factors of water resources and investment which affect the crop yield per unit area, the feed conversion rate, the selection problem between whether to import livestock produce or feed, and the hypotheses for the changes in GDP and population are also taken as factors where large differences in forecast volumes arise.

According to the United States Department of Agriculture's most recent forecast results published in 2007, China's food net-import volume for 2016 would rise to 65.6 million tonnes. Within that, the net-import volume for grain would be 9.19 million tonnes, and the net-import volume for soybeans would be 56.9 million tonnes. In other words, the amount of food lacking would primarily be soybeans (Table 4.3).

Putting together the above forecast results, they are all in agreement that China will become a net-importer regarding its future food supply and demand. However, while they have in common the fact that China has no other choice but to rely on imports in order to achieve a balance in food supply and demand, there are great differences in the forecast figures for net-import volume.

The causes for these forecast results differing are based on: firstly that from 1996 China's cultivated area increased from 96 million ha to 130 million ha; that the proportion of land which can be converted to cultivation use in the future in China

Table 4.3 Previous research on forecasts of China's food supply and demand. Unit: million tonnes

	Supply			Demand			Trade volume (import volume)		
Year	2010	2020	2030	2010	2020	2030	2010	2020	2030
International research									
US department of agriculture	451.2	502.6	–	480.1	563.1	–	28.8	60.5	–
Brown	322.5	299.7	275.6	473.6	558.2	649.6	151.2	258.4	374
Impact	416.7	448.9	–	450	490	–	33.3	41.1	–
Huang et al.	486	570	–	512	594	–	26	24	–
Nyberg/GTAP	–	661	–	–	727	–	–	66	–
World Bank	483.5	–	–	501.8	–	–	18.3	–	–
Overseas Economic Cooperation Fund (OECF), Japan	500.3	–	–	628.1	–	–	127.8	–	–
Major domestic research									
Economic research institute, state development planning commission	–	–	–	–	–	–	–	–	63
Chinese academy of agricultural sciences	–	–	–	–	–	–	20	33	–
National conditions investigation group, Chinese academy of sciences	–	–	–	–	–	–	50	–	–

Source: Compiled from: Qu (1997); Guo (1997); Fan and Sombilla (1997); and Barney et al. (1999)

has been lower than expected; and the necessity of further raising the efficiency rate for the utilization of cultivated land. This became the cause for the United States Department of Agriculture's 2007 forecast that China's 2016 net-export volume of grain would go below 10 million tonnes. In addition, as China's statistical data for livestock produce production volumes has changed greatly, in the 1998 edition of the *China Statistical Yearbook* the 1996 meat production volume was revised downwards by 22.3 %. In the aforementioned forecasts, because this was something that mostly couldn't be envisaged, it was not possible to make projections accurately. Furthermore, for these forecasts, because there was no examination of the selection problem between whether to import livestock produce or feed, and what's more, because the production efficiency rates for livestock produce within and without the country differ, the part of the forecast relating to feed has brought about a large disparity with the whole.

3.2 Results of Forecast for Food Supply and Demand Balance

Otherwise, there is the "Food Supply and Demand Model by Commodity" which Chen developed using a partial equilibrium analysis (Chen 2004). Looking at the forecast results from this model's basic concept proposal, with an assumption of maintaining a per capita GDP growth rate above 7 %, the production and consumption volumes of food in China in the future will increase further, but the imbalance of food supply and demand (production total – consumption volume) has also continuously grown, and will probably increase to 48.83 million tonnes in 2015, and 77.61 million tonnes in 2020 (Table 4.4).

3.3 Results of Forecast for Totals of Four Major Food Commodities

Looking at the forecast results from the above model's basic concept proposal, the total food production for the four major commodities of rice, wheat, maize and soybeans will continually increase, and the consumption volume will also increase without interruption. Looking at the breakdown of the consumption volume, the staple-food-use consumption volume for foodstuffs will gradually decrease,

Table 4.4 Results of forecast for China's food supply and demand balance (basic concept proposal). Unit: 10,000 tonnes

Year	Production volume	Consumption volume	Production volume – consumption volume
2005	48,403	50,376	−1,974
2006	47,581	50,960	−3,379
2007	8,421	51,298	−2,878
2008	48,660	51,366	−2,706
2009	48,731	51,691	−2,960
2010	48,633	51,926	−3,293
2011	48,993	52,445	−3,453
2012	49,109	52,836	−3,727
2013	49,268	53,342	−4,075
2014	49,360	53,815	−4,454
2015	49,507	54,390	−4,883
2016	49,590	54,930	−5,340
2017	49,708	55,552	−5,844
2018	49,778	56,177	−6,400
2019	49,854	56,893	−7,039
2020	49,853	57,614	−7,761

Source: Chen (2004)

but the feed-use consumption volume for foodstuffs will continually increase, and the industrial-use food consumption volume will increase continuously and greatly. The volume for use as seeds will basically be stable, and the consumption volume for other things, such as wastage, will continue to increase (Table 4.5).

Via the above analysis, because the unbalanced parts of China's future food supply and demand will still be large, China's food self-sufficiency ratio will fall to 84 % in 2020, and will fall far short of the policy target which China has raised to "secure the food self-sufficiency ratio at over 95 %". Therefore, in the future one cannot be very optimistic about China's food supply and demand.

4 Examination of Future Food Security Strategy

With the assumption that the Chinese economy has been continually growing, China's industrialization and urbanization have evolved without interruption, and the usable area under cultivation is gradually decreasing. If it relies only on the current level of technology, the probability for future increased food production is low. In addition, for food production there is also the fear of being affected by frequent drought disasters accompanying global warming. Through these uncertain factors, in addition to the shortfall in the food supply seen from the aforementioned forecast results, there comes about a further increase in imports, and there is the possibility of a major impact on international food markets. Consequently, with economic globalization, how to establish China's future food security strategy will become a more important policy issue. Meanwhile, in regions in China such as Heilongjiang Province, Inner Mongolia and Xinjiang, there is still the potential to raise the production per unit area, and they can harness the potential of agricultural land through such things as multi-cropping and improvement of agricultural land. Furthermore, although the quality of China's agricultural land is still low, there is room for improvement. There is also a good deal of potential to reduce damage, reducing natural and biological disasters.

Based on the above analysis, the following matters will be important in looking at China's future food security problems.

First, regarding the food self-sufficiency ratio, it is necessary to make the proper adjustments, assessing the situation for the international food supply and demand balance. Fulfilling 100 % the food self-sufficiency ratio in the future is already unrealistic. The setting of the food self-sufficiency ratio at 80–95 % as a specific policy objective is desirable.

Second, via measures including the development and dissemination of agricultural technology, the putting in place of agricultural infrastructure, and the prevention of disease and insect damage, it is necessary to raise the total utilization efficiency rate of water and land resources and constantly improve the production volume per unit area. Moreover, it is necessary to raise the feed utilization efficiency rate, strengthen aquatic produce and the production of ruminant livestock produce, and develop the food processing industry.

Table 4.5 Results of forecast for the supply and demand of four major food commodities (rice, wheat, maize, and soybeans)

Year	2011	2012	2013	2014	2015	2016	2017	2018	2019	2020
Cultivated area (10,000 ha)	8,503	8,528	8,547	8,565	8,582	8,598	8,612	8,624	8,633	8,639
Yield per unit area (tonnes/ha)	5.224	5.232	5.239	5.247	5.254	5.26	5.265	5.27	5.273	5.276
Production volume (10,000 tonnes)	44,417	44,613	44,779	44,942	45,089	45,223	45,344	45,446	45,525	45,582
Consumption volume (10,000 tonnes)	47,675	48,136	48,621	49,139	49,683	50,251	50,845	51,480	52,173	52,924
Staple food-use (10,000 tonnes)	24,282	24,117	23,945	23,765	23,567	23,349	23,112	22,866	22,623	22,380
Feed-use (10,000 tonnes)	12,434	12,601	12,776	12,962	13,155	13,356	13,566	13,784	14,011	14,247
Industrial-use (10,000 tonnes)	8,499	8,948	9,421	9,927	10,469	11,047	11,664	12,323	13,028	13,783
Seed-use (10,000 tonnes)	975	977	978	979	980	980	981	981	981	981
Wastage, etc. (10,000 tonnes)	1,485	1,493	1,500	1,507	1,513	1,518	1,523	1,527	1,530	1,532
Production volume − Consumption volume (10,000 tonnes)	−3,258	−3,523	−3,842	−4,197	−4,594	−5,028	−5,502	−6,035	−6,648	−7,342
Net-export volume (10,000 tonnes)	−3,723	−3,831	−3,985	−4,175	−4,395	−4,681	−5,133	−5,888	−6,850	−7,774
Change in stored volume (10,000 tonnes)	465	309	143	−23	−199	−347	−369	−147	202	432
Stored volume (10,000 tonnes)	19,399	19,193	18,828	18,300	17,597	16,747	15,877	15,231	14,935	14,870

Source: Chen (2004)

Third is to develop the futures market in focused fashion, establishing food distribution centers domestically. At the same time, establishing price changes via market mechanisms is necessary. Moreover, in addition to taking a series of countermeasures domestically, the implementing of a multifactorial import strategy for agricultural produce is also important. Furthermore, the aiming at the establishment of an effective Northeast Asian food security system is called for, by way of coordinating with Northeast Asian net-food-importing countries, seeking possibilities for the formation of international food clusters, and the constructing of a strategic food-stockpiling system for this region and a dissemination system for agricultural technology.

This paper is a reconfiguration based on the paper by Chen published in ERINA Report Vol. 80 issued in March 2008 and the presentation by Nie published in ERINA Report Vol. 81 issued in May 2008.

References

Barney GO, Bogdonoff P, Qu W (1999) Chinese and global food security to 2030: reducing the uncertainties. Millennium Institute Professional paper #16, 27 Feb 1999

Brown LR (1995) Who will feed China?: wake-up call for a small planet. W. W. Norton, New York

Chen Y (2004) China's food supply and demand and forecasts [Zhongguo shiwu gongqiu yu yuce]. China Agriculture Press, Beijing [in Chinese]

Fan S, Sombilla LQ (1997) The disparities in China's future food supply and demand forecasts [Zhongguo weilai liangshi gongqiu yuce de chaju]. China Rural Survey [Zhongguo nongcun guancha] 3:19–25 [in Chinese]

Guo S (1997) China's food supply and demand and international trade [Zhongguo liangshi gongqiu yu guoji maoyi]. China Rural Survey [Zhongguo nongcun guancha] 3:26–30 [in Chinese]

Qu W (1997) A comparison framework of seven China agriculture models. Millennium Institute workshop paper, 29 Sept

Xu D (1981) China's modern agricultural production and trade statistics [Zhongguo jindai nongye shengchan ji maoyi tongji ziliao]. Shanghai People's Publishing House, Shanghai [in Chinese]

Yu L, Teng Z (2000) General history of China [Zhongguo tongshi] (final volume). Shandong People's Publishing House, Shandong [in Chinese]

Chapter 5
Korea's Food Security Schemes

Jaehyeon Lee

Abstract This article aims to describe the features and issues of Korea's efforts on food security. To approach this aim, trend of food self-sufficiency and schemes for food security as the current agricultural policies in Korea were examined. Accordingly, the article find out three main points as follows; first, the food situation of Korea represented by heavy import reliance and repeating food crises has been forced into low food self-sufficiency rate and instability of food balance still now. Secondly, therefore recently the government established measures for food security such as like target of self-sufficiency rate setting, effective use of stockpiling system and foreign agricultural development. Thirdly, however Korea's policy measures on food security are not coordinated enough to be called a policy package could be act to deal with considerable ongoing uncertainty of worldwide food trade. Thus, Korea needs to recognize the risk, i.e., existence of uncertainty that hinders a stable supply of food, and to reconsider its food strategy in order to implement systematic and pragmatic measures.

Keywords Korea • Food security • Food self-sufficiency • Policy measures for food security • Uncertainty of food trade

1 Introduction

The "Framework Act on Agriculture and Fisheries, Rural Communities and Food Industry" of the Republic of Korea states clearly that its purpose is to secure a stable supply of food for the public (Article 1). The Framework Act sets the target self-sufficiency rates for major food items and requires the government to make efforts to achieve them (Article 14). The government is taking food self-sufficiency as an important item on its policy agenda, due to the country's heavy reliance on importing food, with poor international competitiveness, caused by the small scale of the farming system.

J. Lee (✉)
Faculty of Agriculture, Kagoshima University, Kagoshima, Japan
e-mail: lee@agri.kagoshima-u.ac.jp

In addition to measures to maintain domestic agricultural productive capacity, the government has put in place a distribution system for farm products, a stockpiling system for important food items, and border measures on food trade. These systems have a direct impact on the degree to which food security is achieved. However, Korea's policy measures on food security are not sufficiently coordinated to be called a policy package. Also, it is only recently that the government has become interested in a multifaceted approach to improve food security, such as higher self-sufficiency for grains other than rice or comprehensive emergency measures to cope with, for example, a rapid rise in grain prices or the export controls of producing countries.[1] The following sections describe the features of and issues for Korea's efforts on food security. To start with, an overview of the country's food supply which covers the demand–supply structures of major grains and their self-sufficiency rates is in order.

2 Food Situation

2.1 Food Self-Sufficiency Rate

Korea's Food Balance Sheet indicates the country's calorie-based food self-sufficiency rate stands at 44.3 % at present (FY 2008). The calorie-based food self-sufficiency rate has declined steadily since 1970 and went below 50 % for the first time in 2000. It has not shown any sign of recovery. The low rate derives mostly from the fact that the country relies heavily on imported food. The low self-sufficiency comes from, more than anything else, high shares of imported feed grain as well as oils and fats. However, rice – a staple food in Korea – enjoys high self-sufficiency, which is why the public has difficulty recognizing the fact that the country's overall food self-sufficiency is low. Korea's food situation is almost the same as that of Japan in this particular aspect.

2.2 Demand/Supply Situation and Price of Rice

Rice is Korea's staple food. Therefore, the country's agricultural policy has focused steadfastly on rice. How to increase rice production was the priority issue in the past when rice was in short supply. It is still fresh in our memory that an IR hybrid variety called "Tongil" [unification] was enforced as part of the yield increase drive. Eventually, the country achieved self-sufficiency in rice. However, a stable supply was not secured, and harvests were particularly poor in 1980 and 1995. There are still causes for concern regarding rice supply (Fig. 5.1).

---

[1] Korea Rural Economic Institute (2008) and Kim et al. (2008) are representative examples. These studies were performed after grain prices began shooting up in the global market from 2007 onward.

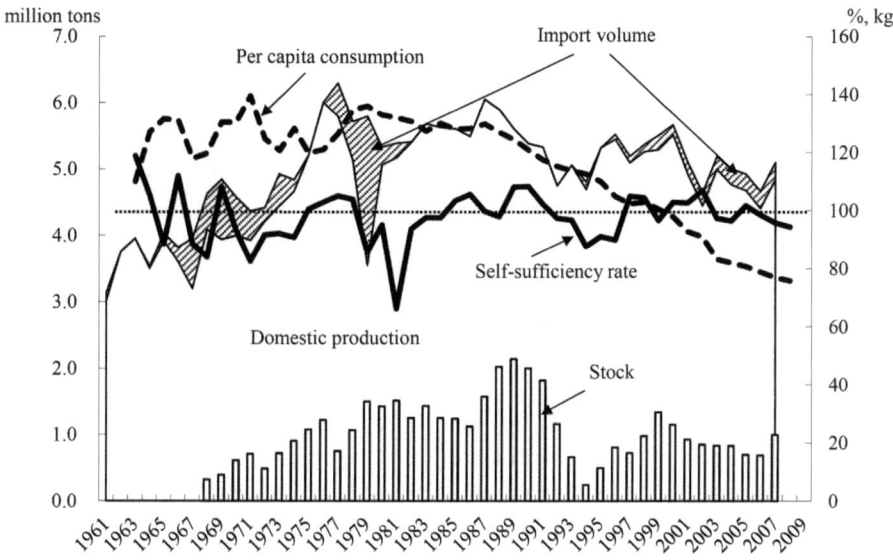

Fig. 5.1 Changes in supply and demand of rice in Korea (1961–2009) (Source: Ministry of Agriculture, Food and Rural Affairs, Agriculture, Forestry, Fisheries and Food Statistical Yearbook, 2009)

In the past, the supply and demand equilibrium was at 5.5 million tons in Korea's rice market. It has dropped significantly to less than five million tons in recent years. The drop was due mainly to the decline in rice consumption which peaked in 1980 and the resultant diminishing of paddy fields.[2] Per capita rice consumption per year in Korea went below 100 kg in 2000. The pace of decline accelerated, and it went further downward to less than 80 kg from 2006 onward (Fig. 5.1). Against the backdrop of domestic rice demand constantly falling, increasing rice imports with the implementation of minimum market access (MMA) have been pushing rice prices downward, and, as a result, the productive capacity for rice may shrink further.[3]

The Korean government abolished the rice purchase system when it reviewed rice policy in 2004. Subsequently, the rice price fell as consumption declined and rice imports rose. Rice farmers' selling price fell from 158,632 won/80 kg (2003 harvest) to 142,720 won/80 kg (2007 harvest).

---

[2] The acreage of rice cultivation has declined steadily since it hit a peak of 1.24 million hectares in 1990. It went as low as 0.928 million hectares in 2008, which is about 75 % of the peak.

[3] Agreement was reached in 2004 to extend the grace period for tariffication as a result of the renegotiation of the opening of Korea's rice market. Accordingly, the following commitments were made. (1) The grace period for tariffication will be extended (for another 10 years, from 2005 to 2014 inclusive). (2) The import quota will start at 225,575 tons in 2005 (equivalent to 4.4 % of domestic consumption) and rise to 408,700 tons (7.96 %) in the final year. (5) State-run trade will be maintained. (6) Part of the imported rice will be sold on the consumer market (starting with 10 % of imports in 2005 and rising to 30 % in 2010). (7) Part of the imports (205,000 tons) will be provided as a quota to designated countries (China, the United States, Thailand, and Australia). (Ministry of Agriculture, Press Release on the Outcome of Renegotiation on Rice, December 2004).

2.3 Import Reliance for Major Grains and Repeated Food Crises

Korea's food self-sufficiency is low primarily because the country relies heavily on imports for oils and fats as well as grains other than rice, such as feed grains. For example, self-sufficiency of grains was estimated at 26.2 % in 2008. If feed grains are excluded, the rate rises to 49.2 %.

Table 5.1 shows the supply and demand for major grains by supply source and use. The self-sufficiency rates for maize and soybeans are below 1 %, and a little less than 80 % of imported maize and soybeans are used as feedstuffs. Where supplied domestically, 77.9 % of maize and 74.7 % of soybeans become feed grain.

Dependence on imported feed grain was one factor which led to food crises in the past. One such crisis occurred in 1997 when feed grain prices shot up. Korea had to devalue its currency (won) significantly in the midst of a currency crisis. Consequently, the price of imported grains soared. At the time, a great number of cattle farmers were unable to obtain enough feed for their livestock. As a result, some small scale farmers lost cattle to starvation, as was widely covered by the media.

Table 5.1 Supply and demand structure and self-sufficiency rates for major grains (FY 2008, 1,000 tons)

	Total	Rice	Barley	Wheat	Maize	Soybeans	Potatoes	Other grains
Carry-in	1,977	695	277	439	506	40	–	20
Production	5,015	4,408	143	9	84	114	228	29
Imports	14,098	243	202	2,373	9,418	1,522	4	336
For feed (%)	62.7	0.0	14.9	9.0	78.4	78.1	0.0	7.7
Supply total	21,090	5,346	622	2,821	10,008	1,676	232	385
Food	5,571	3,755	54	1,628	79	99	111	25
Processing	3,900	655	257	501	1,946	296	63	182
Feed (B)	8,974	0	64	406	7,260	1,194	23	27
Others	701	261	21	30	33	10	35	
Demand total (A)	19,146	4,671	396	2,565	9,318	1,599	232	365
Carry-over	1,944	675	226	256	690	77	–	20
Consumption per capita (kg)	131.4	75.8	1.1	33.7	5.0	8.0	3.6	4.2
Grain self-sufficiency rate (%)	26.2	94.4	36.1	0.4	0.9	7.1	98.5	7.7
Food self-sufficiency (%)	49.2	94.4	43.0	0.4	4.1	28.2	109.2	8.3
B/A	46.9	0.0	16.2	15.8	77.9	74.7	9.9	7.4

Source: Ministry of Agriculture, Food and Rural Affairs, Agriculture, Forestry, Fisheries and Food Statistical Yearbook, 2009

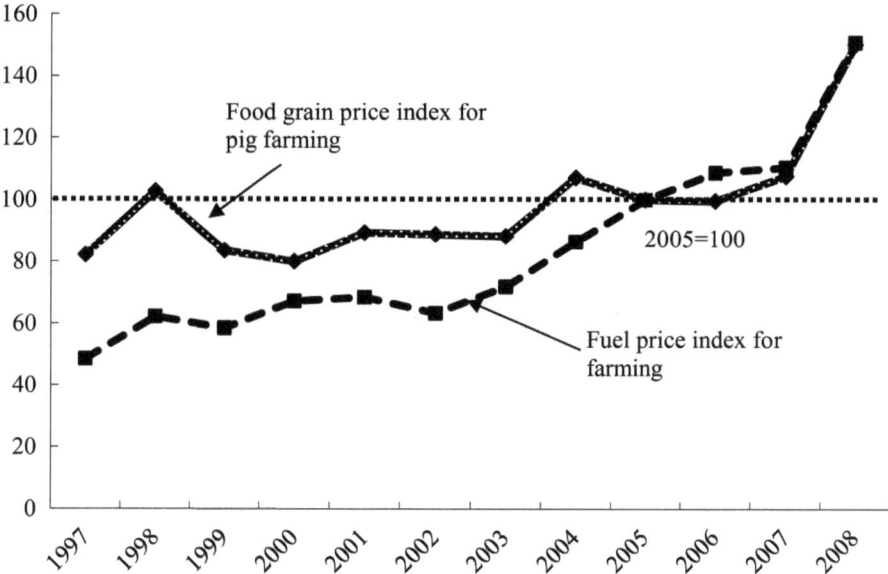

Fig. 5.2 Feed grain and fuel price indexes (2005=100) (Source: Ministry of Agriculture, Food and Rural Affairs, Agriculture, Forestry, Fisheries and Food Statistical Yearbook, 2009)

Another crisis broke out in 2007 when the United States' government changed its energy policy and encouraged the public to shift from fossil fuels to biofuels made from grain. The move destroyed the supply–demand balance for grain and pushed up grain prices in international markets. Compounding the damage were soaring oil prices. Higher oil prices meant price rises in fuel, fertilizers and agricultural chemicals made from oil. As a result, Korea's grain imports declined 2 % in FY 2007, yet the import price was 80 % higher than a year earlier. Figure 5.2 shows the spikes in feed and materials prices in 1997, as well as 2007 onward.

3 Schemes for Food Security

3.1 Stockpiling System

The Korean government stopped purchasing rice in 2004. The system is a typical example of the price support designated by WTO's Agreement on Agriculture because the government fills the gap between the purchase and the selling prices. There was no alternative but to abolish the system. While the WTO pact required reduction of the Aggregate Measurement of Support (AMS), the Korean government spent 90 % of the AMS to maintain the purchase system. Subsequently, price support was replaced by direct payments, such as the rice farmer compensation system. Also introduced was the rice reserve system where the government purchases a fixed amount of rice and keeps it as a reserve. The purpose of the system is to assure food security. The reserved rice will be released to the public at the time of bad harvests or other emergencies. Under the system, the government purchases 432,000 tons of rice and stores it. The volume is equivalent to one month of domestic consumption.

The Korean government also has a reserve system for potatoes and condiment vegetables, such as garlic and onions. But the system is seldom put into action, and the stock is very small, if held at all. The non-rice food reserve system is more a kind of price stabilization measure than a food reserve system for assuring food security. The measure has a role of preventing price fluctuations occurring from oversupply.

3.2 Target of Food Self-Sufficiency Rate

The "Framework Act on Agriculture and Fisheries, Rural Communities and Food Industry" requires the government to set targets for self-sufficiency (Article 14–2). Table 5.2 shows the item-by-item and category-by-category self-sufficiency targets worked out by the Food Self-Sufficiency Rate Advisory Committee.

The self-sufficiency target in calories was set at 47 % for 2015, a rise of 0.3 % from 46.7 % in 2004. Table 5.2 anticipates that the self-sufficiency rates for almost

Table 5.2 Korea's food self-sufficiency targets (2015)

Item	2003	2004 (A)	2005 (B)	B–A
Rice (A)	97.4	96.5	90	△6.5
Wheat, etc. (B)	7	7.6	4	△3.6
Staple self-sufficiency rate (C)=A+B	68.2	65.3	54	△11.3
Beans (D)	29	25	42	17.0
Yams and potatoes (E)	109.1	107.6	99	△8.6
Grain self-sufficiency (F)=C+D+E+Feed grain	27.8	26.8	25	△1.8
Roughage self-sufficiency rate	84	83.1	85	1.9
Vegetables	94.6	94.3	85	△9.3
Fruits	85	85.2	66	△19.2
Milk and dairy products	80	73	65	△8.0
Meat	70.8	79.3	71	△8.3
Beef	36.3	44.2	46	1.8
Pork	93	86.9	81	△5.9
Chicken meat	76.3	90	80	△10.0
Chicken eggs, etc.	100	100	100	0.0
Self-sufficiency rate in calories	45.6	46.7	47	0.3

Source: The deliberation council for food policies, A proposal of targets for self-sufficiency of food and staples, 2007

all items will decrease. The targets are conservative ones in which the government is ready to accept decreasing self-sufficiency. On the other hand, the government has set achievable targets.

Meanwhile, self-sufficiency is predicted to rise in beans, coarse feed and beef. Domestic production of beans hit a low and turned upward in 2004. The government has probably taken it as a sign of recovery, which is why it has raised the self-sufficiency target. One factor that helped push up the self-sufficiency target for soybeans is the rising demand for domestically produced soybeans as consumers object to genetically modified (GM) soybeans.

Feed-grain farming and coarse feed supply are expected to expand thanks to the Coarse Feed Production Improvement Project, which is stimulating an increase in the self-sufficiency rates for coarse feed and beef. Incidentally, coarse feed supply peaked in 1995 at 7,602 tons. It declined steadily and hit a low in 2000. Then, it began rising again and reached 5,054 tons in 2008, which is a significant rise compared to the 3,392 tons in 2000. The self-sufficiency rate of coarse feed is maintained at around 83 %, largely because of stipulations within incentive packages including cash payments to promote cultivation of feed grain in rice paddies.

3.3 Border Measures and Overseas Agricultural Development

3.3.1 Border Measures on Market Access

Korea has the status of a developing country under the WTO Agreement on Agriculture and maintains preferential measures on rice, high tariffs on certain items and a tariff-rate quota (TRQ) system on many items. Thus, it provides relatively supportive border measures. It cannot be denied that these border measures push up the import prices of agricultural products in the domestic market, and thus hinder consumers' food accessibility. However, at the same time, these are necessary measures to protect domestic farmers who are uncompetitive in the global market. In that sense, the border measures are part of the country's food security strategy. Therefore, the Korean government is arguing for the continuation of developing country status and rejecting a large reduction of tariffs or fewer TRQ items in the ongoing new round of WTO negotiations.

3.3.2 Overseas Agricultural Development

In Korea, few trading companies do business in international grain markets, and they prefer spot transactions rather than forward contracts. Therefore, diversification of supply sources including overseas farms and transaction methods is an important factor in securing steady food imports.[4]

---

[4] Many of the facts about Korea's food import process and overseas farm development are confirmed by the results of the survey and analysis of Kim et al. (2008).

Korea places importance on "overseas agricultural development" as a means to keep major grains constantly entering the domestic market. The "Framework Act on Agriculture and Fisheries, Rural Communities and Food Industry" states in Article 58 that "the government shall support the study of foreign investment in agriculture, fisheries and related industries as well as farm/fishing businesses' foreign investment. In addition, the government shall implement measures necessary to secure overseas fishing grounds." In response, many Korean agricultural organizations and private companies have gone overseas in recent years to set up large-scale farms in places such as Russia, China, and Indonesia. Such agricultural organizations and companies are operating large farms ranging from tens of thousands to millions of hectares and growing food crops such as rice, soybeans, wheat, and potatoes. It is noteworthy, however, that most of these operations have failed and have had to close down, according to related research.

Nonetheless, the Korean government regards developing large overseas farms as an effective tool to maintain constant imports of major grains and to develop overseas markets. The government expects that the overseas farms will sell products to the foreign market in peacetime, and, in emergencies, they will preferentially export major crops to Korea. For that purpose, the Korean government is playing an important role in building partnerships with the host countries, undertaking research to gauge how much demand there is for such business, and collecting and supplying a variety of necessary data.[5]

3.4 Incentive for the Growth of Domestic Agriculture and the Maintenance of Production Base

Korea's Farm Bill (FY 2010) focuses on two basic principles. Firstly, great emphasis is placed on strengthening future growth drivers to develop a small but strong agri-food industry (Ministry of Agriculture, Food and Rural Affairs 2010). The second is a reconsideration of agricultural subsidies, and efficient and flexible use of fiscal measures such as reform in the spending structure through consolidation of similar projects, etc. Although it is important to make domestic agriculture more competitive, government policy on agriculture and the food industry places the highest priority on making agriculture an attractive industry through building a food chain in which the food industry participates. In order to achieve that goal, the government is putting certain policies into effect on different fronts: scaling-up of farms, laying the foundations for producing and distributing safe and high-quality farm products, research and development on farm products and food items that can compete in the global market, and building a food industry cluster.[6]

---

[5] Agricultural organizations and the private sector are supported by the Korea Rural Community Corporation in their overseas farm development efforts.

[6] For more information, refer to Fig. 15.1 in Lee Jaehyeon: The Trends and Potential for Food Industry Clusters in Korea, in this book.

Meanwhile, the direct payment system is being improved in Korea to cope with the reduction in domestic subsidies (AMS). Incidentally, the current system has five types of direct payment: (1) the Rice Income Compensation Direct Payment; (2) the Environment-friendly Farming Direct Payment; (3) the Less-favored Region Direct Payment; (4) the Landscape Preservation Direct Payment; and (5) the Management Succession Direct Payment. The system plays an important role in food security in that the above, excluding the Management Succession Direct Payment, are aimed at assuring farmers a stable income and maintaining a proper level of domestic agricultural production.

The Agricultural Production Base Improvement Project has a certain share in the budget allocated for agriculture (16.6 % of total spending for agriculture, forestry and fisheries) as it has a direct impact on the obtaining and maintenance of farmland. Major items in the project are irrigation, farmland improvement works, and development of large farmlands. It actually has the world's largest farmland development project: the Saemangeum Reclamation Project.[7] Japanese agriculture faces the aging of farmers and the depopulation of rural communities. Unlike Japan where neglected farmland is a pressing issue, it is noteworthy that Korea is actively developing new farmland.

4 Features and Challenges in the Strategic Approach on Food Security

4.1 Issues for Korea's Food Security

Food self-sufficiency may not matter as long as food is constantly coming from different countries and always accessible in whatever quantity. Another issue is the level of food reserves that need to be maintained. These are questions that weigh heavily when we think about what is the necessary level of self-sufficiency, how much of the production base needs to be maintained in the country and in what ways we need to import food from abroad.

Another factor that defines our food policy is the direction in which ongoing WTO agricultural negotiations are heading. Given the possibility of Korea having to give up its status as a developing nation and to reduce tariffs significantly on key items as a result of the WTO's new round of talks, setting the self-sufficiency target prematurely at this juncture would raise doubt about its feasibility. Considering the North–South division of the Korean nation, a long-term strategy for food security

---

[7] The project began in 1991. Reclamation of the western coast of North Jeolla Province will add 28,300 ha of new land, 8,580 ha of which will be developed as farmland. For more information, visit the web site of Saemangeum Project Office at http://www.isaemangeum.co.kr.

needs to take unification into account. However, doing so is not easy because it depends on political decisions to be made in the future. Currently, structural reforms are required for designing a large-scale farming system to increase the productivity by the small-scale farming system based on family farms. Under the circumstances, income compensation and other forms of direct payment may eventually weaken the potential of domestic agriculture since it works as a disincentive for farmers' efforts to raise productivity and efficiency.[8]

4.2 Features of Food Security Strategy in Policy Deployment

The most conspicuous feature of Korea's food policy is the lack of a policy framework in domestic policy. At the moment, the country has neither an official opinion about food security nor definition of and countermeasures against contingencies that hinder a stable food supply. Only the rice stockpiling system is given the status of a means of food security. However, as has been stated, Korea has a number of policy measures in practice that aim to maintain the domestic agricultural base at a certain level of self-sufficiency and to provide stable food imports. In that sense, the country still lacks a comprehensive policy framework worthy of being called a Food Security Strategy.

Secondly, as far as domestic measures are concerned, priority is placed on, if anything, strengthening farm management to revitalize domestic agriculture and the food industry, and on developing export-oriented agriculture and food industries, rather than on improving self-sufficiency or on reserving food for emergencies.

Thirdly, building a mechanism of stable imports is part of the food security strategy. The mechanism tackles the process from the farms in the exporting country down to import procedures. The strategy states that domestic importing companies need to take the initiative in doing business in the global market, and the necessity of having many different import routes. It also specifies transaction methods. Furthermore, it shows a great interest in the kind of overseas agricultural development in which domestic companies take the lead in running large-scale farms in exporting countries.

In summary, the country is not only raising self-sufficiency through a reinforced production base and maintaining border measures, but also adapting to the globalization of food markets and developing overseas farms and export markets. In short, it maintains a flexible attitude toward food security.

---

[8] For more on this, refer to Lee (2006a).

Conclusion

Korea's fragile agricultural structure presents a range of challenges such as an increasingly shrinking and aging farming population, delayed structural reform, and uncompetitive domestic farm products priced far higher than imports.[9] In the event of the failure of the global strategy, such as in developing export markets and overseas farms, large investments by the government and private sector will not lead to recovery and stable imports will be difficult. If domestically produced farm products are found to be no different from imports in quality and safety, imported food will expand its share in domestic markets. If these concerns become a reality, self-sufficiency will inevitably decline further and government will have difficulty in responding proactively to emergencies. The country needs to recognize the risk, i.e., the existence of uncertainty that hinders a stable supply of food, and to reconsider its food security strategy in order to implement systematic and pragmatic measures. When rethinking the strategy, it is necessary to take into account the kind of policy items that will form the core of the new food security strategy: expanded stockpiling systems to cover important food crops in addition to rice; food distribution controls to prepare for severe shortages; and strengthening international partnerships on food procurement.

References

Kim M et al (2008) Occurrence possibility and countermeasures against food security problems (C2008- 28). Korea Rural Economic Institute, Seoul [in Korean]
Korea Rural Economic Institute (2008) Long-term strategy and implementation plan of Overseas Agricultural Development. Study report commissioned by the Ministry for Food, Agriculture and Forestry and Fisheries, Seoul [in Korean]
Lee J (2006a) Features of rice farming structure and issues of direct payment in Korea. In: Kishi Y (ed) Direct payment systems of the world. Agriculture and Forestry Statistics Association, Tokyo, pp 186–199 [in Japanese]
Lee J (2006b) Environmental changes and problems facing agriculture in Korea – focus on paddy-field farms and rice market. J Rural Issues 59:10–20, Tokyo [in Japanese]
Ministry of Agriculture, Food and Rural Affairs (2010) Outline of FY2010 budget and fund management plan, Seoul [in Korean]

---

[9] Lee (2006b) provides a detailed account of the features and structure of Korea's agriculture.

Part II
Food Clustering in Northeast Asia

Chapter 6
The Food System Based on Agriculture

Food Collaboration

Osamu Saito

Abstract Food industry cluster and local branding are regarded as effective strategies for local revitalization. Small and medium-sized food and related businesses from inside and outside a community are brought together in producing areas to form the food industry cluster and competitiveness enhancement for both agriculture and food industry is expected from a strategic partnership between them as they possess different kinds of management resources from one another. The food industry cluster realizes technological innovations, which leads to the development of new industries and products through mutual provision of management resources and technology transfers. Such clustering also contributes to the establishment of local brands by utilizing local resources for the creation of more jobs and yielding income growth to the community. On the other hand, Sextiary Sector (the Farmer's Complex) business is an approach which puts emphasis on integrating mid- and downstream businesses from an individual farm producer's standpoint. Therefore, an integrated approach which takes advantages of both the food industry cluster and Sextiary Sector characteristics is required to create added values inside the community, bringing about more income to producers and further local revitalization through the promotion of job creations and effective use of local resources.

Keywords Food industry cluster • Sixth-Order Industrialization • Local brand

1 Subject Setting

As local economies are in an unprecedented crisis, many factories in farming communities are either being closed or relocated, and small-scale commercial and industrial activities are apparently on the decline. The oligopoly of the food industry

---

O. Saito (✉)
Food and Resource Economics Course, Chiba University, Chiba, Japan
e-mail: osaito@faculty.chiba-u.jp

© Springer Japan 2016
L. Kiminami, T. Nakamura (eds.), *Food Security and Industrial Clustering in Northeast Asia*, New Frontiers in Regional Science: Asian Perspectives 6, DOI 10.1007/978-4-431-55282-6_6

first started in the food processing business. Small and medium-sized businesses based in rural areas have lost competitiveness and have been struggling for survival. As the 1990s unfolded, new mid-stream and downstream business models appeared from those rural communities that were uncompetitive as producing areas, and value chains have been formed. Such value chains took in different downstream businesses (farmers' stores, restaurants, etc.) and moved closer to consumers. A new food system emerged which covers production all the way through to consumption and seeks networking with many outsiders. Such networking is critical in building partnerships with the food industry to make up for the management resources which are scarce in rural areas. Even for those rural areas that have concentrated their management resources on agricultural production, there are plenty of business opportunities up for grabs to integrate the mid- and downstream positions in the food chain. Food companies promoting themselves in agricultural producing areas now look not only for the procurement of raw materials and foodstuffs but also for various resources that exist in rural areas.

It was necessary for producing areas to attract processing factories to raise productivity, let innovation take place and develop new businesses so that area-wide sales will rise and new jobs will be created. At the same time, in order to exploit the local resources more effectively, it was increasingly important to form a food industry cluster to make the agriculture and food industries more competitive. Under the circumstances of the building of clusters, which have economies of scale and synergistic effects and are participated in by small and medium companies in rural areas, strategies aiming at local revitalization have strengthened the relationship between agriculture and food industries.

Among other farm products, pickled plums produced in Wakayama Prefecture are regarded as the typical food cluster model. It has a solid position in the domestic market share, although imports have a 60 % share. It also functions as a basis of competitiveness as it sources the primary processing to farmers, develops the marketing channels, works with businesses to bring about innovation in product and resource development, uses local resources and creates jobs. The wine industry is another example. Imports already have more than half of the market share. Efforts are being made to form partnerships with agriculture through such measures as quality improvement to make domestic wine more competitive, support it with a protected designation of origin system, the introduction of new varieties, innovation in cultivation method, and brand building for domestic wine.

Study of food industry clusters and local brands is spreading from Japan to Korea, and China. As globalization and localization are going forward in parallel, many issues need to be dealt with: changes of distribution system; joining supply chains with value chains; utilization of local resources through agriculture–commerce–industry collaboration; and establishment of a food system (Yagi 2008). Under the circumstances, attention is focused on the food industry cluster, business diversification and local branding, as these are regarded as an effective strategy for local revitalization. Expectations are mounting for local brands to become an effective tool in managing intellectual property, cope with large retailers' PBs, building consumer trust, raising quality control levels, spurring reevaluation of local resources and enhancing their local brand image. For small businesses in the community, competitiveness comes from better utilization of local resources, building a local brand rather than a company brand and quality control improvement. The local brand concept began with the AOC, PDO and PGI of France and the EU and subsequently expanded into intellectual property control and brand asset management. A local brand usually deals with more stakeholders than a company brand does, which is why a certificate system is necessary to control them.

This section's studies discussed how to combine the strategy of food industry clusters and the "Sixth-Order Industrialization" as the scheme for rural diversification, because the food industry cluster strategy tends to focus on product development projects and has little ripple effect on farmers and the community; the same is true with the *noshoko-renkei* (=agriculture–commerce–industry) collaboration project of the Ministry of Economy, Trade and Industry. Many projects fail to form a strategic alliance between different economic units or to build a comprehensive chain that includes marketing. In regard to the other project relating to rural diversification, the "Sixth-Order Industrialization" should play a greater role in utilizing local resources to create jobs and raise incomes.

2 Food Industry Clusters and Sixth-Order Industrialization

2.1 Issues for Japan and East Asia

The first singularity of Japan is the fact that medium to small food businesses exist over a wide area, play a significant role in creating added value and jobs and are in a good position to work with agriculture. Many food-related companies which give the respective domestic and imported raw materials and foodstuffs a proper position are good at utilizing local resources and have built a good relationship with the rural agricultural sector.

Secondly, as the distribution system has changed greatly – but not to the extent of the EU or the US where oligopoly and PBs are dominant – and as competition goes on between systems under the strong influence of the retail sector, the supply chain and value chain need to be integrated in ways to create value to raise efficiency, engage in profit sharing and gain consumer trust. For example, as the agricultural production system is weakening, volume retailers are using "enclosure" tactics to improve quality control for their PBs.

Thirdly, the "Sixth-Order Industrialization" project has been smoothly put into practice with the growth of agricultural corporations. Also, production, processing and marketing processes are being integrated as food-related businesses enter the agricultural sector. As a result, it has become easier for them to utilize each other's resources. In other words, they have come to realize that it is less risky for the collaboration between different industries to use the different sets of management resources they have in a complementary manner to grow their businesses than to integrate investment and knowledge pools.

Fourthly, as deregulation makes it difficult to implement national projects and local administrations play a greater role in platform building and principle/strategy making, budget allocation is dwindling under the slogan of decentralization of power.

Fifthly, the Ministry of Agriculture, Forestry and Fisheries advocated a scheme to promote local food, used food industry clusters as a keyword without developing a good understanding of it, and proposed Sixth-Order Industrialization in relation to the agriculture–commerce–industry collaboration projects of the Ministry of Economy, Trade and Industry.

Although the former two tend to take up product development projects which are easy to support with subsidies, they produce little ripple effect because strategic partnering is lacking. In contrast, the Sixth-Order Industrialization is most likely to generate income in the community. However, technology transfer from outside food and related businesses and innovation are the subjects specific to the food industry cluster.

Sixthly, Japan and Korea have a similar background in regard to food industry clusters and local brands. As food makers in Korea do not have business bases in rural areas, it is difficult for them to form a cluster with the processing industry at its core. So, branding strategy is more often focused on raising brand value and sales promotion than on quality control levels. In Korea, large retailers' PBs and local branding have grown rapidly, but the low quality control levels of local brands make price competition happen more easily at the retail level.

Seventhly, in China, foreign and domestic food companies take the lead and integrate the agriculture sector in a specific way. A good example of this is the practice of owning a directly-managed farm. Unlike in Japan or Korea, the food system in China is neither advanced nor mature, and downstream systems do not compete much. As China is a vast nation, in order to utilize different local resources in branding to compete with imports, they have to have in place systematized techniques to improve delivery, selection, packaging and taste to raise quality control levels. However, high-end items such as Chinese tea or Chinese mitten crab already have a better brand management system and certification system in place.

2.2 Food Industry Cluster Strategy

In Michael Porter's cluster, the system consists of four components – element conditions (input resources), demand conditions, corporate strategy and competitive environment, and related/support industry – and together they make it easier for innovation to take place and thus to make the cluster more competitive. The food industry cluster discussed here produces innovations – development of new industries and products, establishment of local brands, and formation of both value chains and supply chains – so that income and jobs in the community will grow and the local resources will be used more effectively.

The concepts of the food industry cluster that have been devised in Michael Porter's industrial cluster make consideration of the characteristics of food, agriculture and locality. As the cluster evolves, the component units are linked more strongly together. The four conditions that constitute Michael Porter's "diamond" interact and work as a system in the community, which in turn makes technological innovation happen and causes an overall revitalization. Conflicts take place in the community, but coordination capability emerges to solve them. As a competitive cluster is formed and recognized by consumers, it offers an incentive that attracts food company factories from outside. The cluster becomes bigger, and more demand for raw materials and foodstuffs is created. At an early stage, however, the coordinator plays an important role in linking different units.

Information and biotechnology industries are often seen as business models for cluster formation. Food and related industries incorporate traditional industries closely linked with the local community. Thus they tend to have an advanced clustering and procure high percentages of raw materials and foodstuffs from their local rural areas. The related industry contains within it a kind of service industry that forms a cluster of foodstuff and hospitality businesses in sightseeing spots using local agricultural resources, and builds strong ties with agriculture.

The first thing that needs to be said about the concept of industrial clusters is that the distance inside the cluster is the physical proximity between the economic units. Sharing of information and tacit knowledge is easy there, and relations between competition and cooperation continue inside the local community.

Secondly, small and medium-sized businesses having difficulty developing new products or experiencing innovation from their own resources may rely on a partnership or tie-ups with other businesses to secure the management resources they need. Such a partnership of tie-ups is helpful not only in product development but also in procurement, marketing, planning and management, and logistics.

Thirdly, as a cluster is formed in the rural areas and branding makes progress through marketing of individual businesses and supportive policies of the local government, the cost of entry for newcomers goes down, the standard technology accumulated in the community becomes available, and sales channels are established. Also, an industrial cluster draws new players to and causes technological innovation in new business fields.

2.3 *The Sixth-Order Industrialization Strategy*

One of the singularities of the Sixth-Order Industrialization as a system of endogenous agri-business led by rural diversification is the fact that competitiveness comes from its foundation on agricultural production and the distribution of the middle- and downstream positions on the food chain.

Since the mid-1990s, Saito et al. have been emphasizing the necessity of endogenous agri-business led by rural diversification (Saito 1999). This approach is almost the same as the Sixth-Order Industrialization concept and the framework is easy to theorize. Later, the Ministry of Agriculture, Forestry and Fisheries turned to the concept as a tool to promote the local food industry. The ministry followed in the footsteps of the industry cluster concept of the Ministry of Economy, Trade and Industry and adopted the food industry cluster strategy to make innovations happen through industry–academia–government partnerships. Many food-related companies are located in rural areas and they contribute to the sales growth and job creation of local food businesses. In order to revitalize both the food industry and agriculture in the globalized market, a strategy is called for that forms food industry clusters and brings the food industry and agriculture work more closely together. As the agriculture and food industries (not limited to small businesses) have different kinds of management resources, gain in competitiveness is expected from the strategic partnership between them. Although a food industry cluster is usually about a processing of certain products in which a cluster of local businesses are formed, greater importance should be placed on joining with different categories, such as farmers' markets, restaurants, and green tourism, including farm stays. Since "buy local campaigns" are not clearly defined, the "local food system" is more effective as a concept.

The Endogenous Agri-business Concept has been used as a theory because there has been little discussion about Sixth-Order Industrialization. The food cluster concept, on the other hand, has proposed different frameworks such as provision of management resources, technology transfer, product development and innovation through the collaboration of outside food companies and the farming community. Importantly, from the perspective of local revitalization, the Sixth-Order Industrialization and food industry cluster need to be understood in an integrated manner. However, the Sixth-Order Industrialization may place too much emphasis on the integration of production, processing, marketing and collaboration inside the community. In such a case, the advantage of a food industry cluster would be lost because it is supposed to build partnerships with external food companies to make innovation possible. Therefore, the Sixth-Order Sector and food industry cluster strategies would have to be integrated.

The types of food companies that desire collaboration with local firms from outside are likely to have plenty of management resources and be better positioned to take the risk of joining with farmers or a rural community.

The Sixth-Order Industrialization concept of the Ministry of Agriculture, Forestry and Fisheries goes beyond materials to include the energy industry. Each component of the food system, including the consumers, has to take advantage of clustering and vertical collaboration in order to share the benefits and make each other more competitive so as to make matters more efficient.

Local agricultural corporations have formed a value chain – such as farmers' markets, processing facilities, restaurants, or direct sales to consumers – and complemented the fragile production system. However, when integration deepens, more management competence and capital investment are required. It is therefore efficient to put in place a division of labor through food–agriculture collaboration.

2.4 "Noshoko-Renkei" Projects and Product Development

Many "*noshoko-renkei*" projects, as means of agriculture–commerce–industry collaboration promoted by the Ministry of Economy, Trade and Industry are inclined to product development where a couple of economic units are linked vertically (Japan Small Business Research Institute 2009). In many cases, raw material/foodstuff, processor and marketing units tie up in the hope that combination of different management resources will have a synergistic effect on product development. However it is difficult to analyze important issues: whether it is a strategic partnership involving mutual transfer of management resources (knowledge, technology, human resources, capital); or whether the positioning of product development is clarified. Upstream producers seldom play a leading role in product development. When production and processing are integrated, it is always an "agriculture–agriculture collaboration". In some cases, a research institute not only supports but also works with the local food and related businesses and does technical verification to promote the product widely. From a product development point of view, if such collaboration covers a long period of time from development to extension involving high cost, it tends to be a simple study such as functional analysis. If "*noshoko-renkei*" takes place over a couple of years and requires product development costs, a long experimental phase and continuous investment are usually called for. Unlike those cases that use agricultural subsidies, production expansion requires the food and related businesses to make investments themselves. On the contrary, kneading processed vegetables or fruits into bread, noodles or sweets or adding them to liquor do not differentiate the product much, but a certain level of sales is achievable for a relatively low development cost.

In "*noshoko-renkei*" projects, the agriculture unit is often regarded as the procurement source of raw materials rather than a strategic partner. Also, product development has little ripple effect because it neither creates jobs nor distributes income within the community. Product development in a food industry cluster has similar shortcomings. Reorganization of Food Councils across the country into a nationwide organization has produced little effect. Instead, local revitalization would be driven largely by income growth and resource utilization as farmers' markets grow and are integrated with restaurants and processing facilities.

2.5 Self-Sufficiency, Market Size and Food Industry Clusters

Reduction in self-sufficiency changes the relationship between food and related businesses and has a definitive impact on the formation of the domestically produced food system. In general, when the self-sufficiency of a certain item declines, the food and related industries for the item likewise diminish. Consequently, large corporations would have difficulty securing a stable supply domestically, and the medium and small enterprises will then be required to play an important role as a source to procure the foodstuffs. Dwindling supply of domestically sourced materials pushes up the price and brings the procurement volume of food and related corporations down further. Demand for domestically produced food would decline if their price rises to two or three times that of imports sold at large retailers and CVM. When the prices of domestic products rise sharply, high-end items will probably be sold only at department stores and low-end ones at farmers' markets in the rural areas. When sales channels are limited, the market becomes smaller and consumers have fewer places to buy the product. Food is relatively inexpensive at large retailers and convenience stores. If domestically produced food loses these channels, a significant proportion of consumers in urban areas will have difficulty gaining access to it.

If supply decreases and the market shrinks, development of production materials will be delayed and the improvement and enlargement of production systems will be hard to accelerate. If such a vicious cycle continues, domestic food systems will enter stagnation. When the sufficiency rate falls to 10 % or so, large corporations have difficulty securing a stable supply of domestic materials and small businesses in the food market are left out.

3 Implications and Roles of Local Brands

3.1 The Perspective for Local Brands

Local brands include not only farm products but also crafts, hot springs and other service industries. Farm products are relatively inexpensive and consumers often buy them specially. Furthermore it is easier for local brands to exploit local resources to differentiate them from others in safety, taste and function.

Local brands take special care in the production sequence, so that the principal challenges are quality control and brand value improvement. Region-wide brand management is often problematic. In many cases, local brands compete with large retailers' PBs. There are nine singularities (Saito 2007, 2009).

The first is the advantage of having a local brand. Even the small food companies in the local community already have an individual brand. Having a local brand gives them potential advantages such as higher consumer recognition and better quality control through technological innovation. For companies that already have a solid corporate brand, the resulting corporate image and consumer recognition would make it easier for them to market new products. However, if the local brand is not solid enough both in its image and management, sales growth would be hard to come by.

Secondly, if a basic quality control system is not in place, implementing one is more important for farm products than having a good image. In such cases, it is necessary to combine brand elements and to have a brand hierarchy. Introduction of brand hierarchy gives producers an incentive and builds up a mechanism of self-improvement. Business expansion would take place, from delivering fresh products to processing them, thereby pushing up the brand evaluation of both the fresh products and processed goods.

The third point concerns the brand element. Although emphasis is often placed on character or package design, in the case of farm products' safety, taste and function are also missions for brand elements. The protected designation of origin implemented in France and the EU needs to be understood before thinking about the brand value in a marketing context.

The fourth point is public sector support. For farm products, breeding, trademark registration or production patents as a way for branding often require public support such as intellectual property control or certificate systems. So far, priority has been placed on brand protection, and little thought has been given to utilization. As for the Japanese "regionally based collective trademark system" in particular, little discussion has been made on the role of the supply zone and control bodies for local brand management.

The fifth point is brand hierarchy. In order to make the local brand solid, it is necessary to put a brand hierarchy in place. A top brand with a relatively small share can benefit from the local image it is given and a fixed sales channel. Top and middle brands will see a better communication with clients, and find it easier to move forward in the marketing relationship and make two-way proposals.

The sixth point is the coordination between the PB and local brand. When the local brand has a proper hierarchy and positioning, coordination with the PB becomes better and proposals are more effectively made to the retailer.

The seventh point concerns the new entries using the brand. As the local brand gets higher recognition, more businesses join in, and production–processing integration pushes up the level of quality control and efficiency, which in turn raises the brand value and sales.

The eighth point is quality control levels. To what extent local small businesses utilize the local brand depends on the quality control levels of the local brand and their brand value. If the local brand is not up to par in quality control, companies with a good performance will never take part in the local brand.

The ninth point is the relation with the brand umbrella effect. For example, green tourism and eco-tourism are effective tools to raise the brand umbrella effect. In order to raise the brand value, it is necessary to spend time and money on building a good communication channel. However, integration of production and service is not the subject of brand strategy, but the issue of rural diversification.

3.2 Conjunction of Supply Chain and Value Chain

As globalization progresses and the downstream sector of food chains takes the initiative in building a distribution system, a kind of structural transformation takes place in the food system that changes the roles of each economic unit and integrates vertical relations.

In brand building, integration means the conjunction of both a supply chain and a value chain. When a supply chain and value chain are in place, partnerships between large retailers and producers are built, efficiency rises and added value is created at the same time, and a win–win relationship becomes easier to form. More efficient chains make wholesalers and other intermediaries unnecessary, and processing and product development become easier. The supply chain puts in place a traceability system covering the entire process from the farm down to downstream businesses and consumers and a chain of materials–production–processing–retail store/restaurant. The discrete chains that existed between production–materials, production–processing and production–sales/restaurant are increasingly integrated.

Viewed from the aspect of food and related businesses, the producers' support helped them to integrate the material, processing and sales functions and form a value chain. When such combinations become more compact and integrated, either partnership or ownership (direct management) has to be chosen. If the ownership model is chosen, it is necessary to make a larger investment and take risks.

Conclusion

The "Sixth-Order Industrialization" as a measure for a rural diversification concept has seldom been discussed in depth. Instead, the endogenous agri-business creation in rural areas concept has been providing the theory. Although the food industry cluster is based on a different set of theories, such as provision of management resources, technology transfer, product development and innovation, the "Sixth-Order Industrialization" and the food industry cluster need to be understood in an integrated manner in order to achieve local revitalization.

One principle of the "Sixth-Order Industrialization" is that the agriculture and rural aspects hold the core competence and initiative to create jobs and raise income.

If the local community and agricultural producers are not in a position to take the initiative, the strategy needs to focus on involvement of private businesses or on "*noshoko-renkei*" as the agriculture–commerce–industry collaboration. In such a case, a food industry cluster will probably be formed in which the initiative is taken by food and related businesses. Nevertheless, it is necessary to build a mechanism of technology transfer and provision of management resources so that the management units on the agricultural producers' side become independent and a win–win relationship is constructed. If both the agriculture and food-related businesses do not have enough power, the local governments need to take the initiative, and construct a platform and strategy for agriculture–commerce–industry collaboration.

As agricultural products are an important part of a local brand and small-scale processors can benefit from the local brand, the quality of products needs to be improved. In order to compete with large retailers' PBs in the fresh products sector particularly, the producing area needs to raise the level of quality control through brand hierarchy and market positioning to build a good reputation. Building mutual trust and a better communication channel with consumers/clients through good brand management will probably expand the scope of business from transaction to interchange, strengthen the ties with local resources and raise brand values. Also, a solid and strong local brand will attract new businesses.

References

Japan Small Business Research Institute (2009) Study on project development through agriculture–commerce–industry collaboration. Japan Small Business Research Institute, Tokyo [in Japanese]

Saito O (1999) Development of agri-business under local initiative and local revitalization. In: Food system innovation and corporate behavior. Association of Agriculture and Forestry Statistics, Tokyo [in Japanese]

Saito O (2007) Food industry clusters and local brands. Rural Culture Association, Tokyo [in Japanese]

Saito O (2009) Strategic challenges and management system of local brands. J Rural Probl 45(4):324–335 [in Japanese]

Yagi H (ed) (2008) Economic interdependence and agriculture of northeast Asia. University of Tokyo Press, Tokyo [in Japanese]

Chapter 7
The Network Structure of a Soybean Cluster in Hokkaido

Teruya Morishima

Abstract Soybeans are a crop being grown all over the world, mainly for use as oil and as a source of vegetable protein. But domestically produced soybeans in Japan are basically for food use, for example tofu, natto [fermented soybeans] and soy milk. Hokkaido is the largest area of production constituting approximately a quarter of the national total. Therefore in Hokkaido, "Soybean Cluster" was formed from a partnership between the soybean producers, processing companies, government, and academia in the area. The aim was to develop new industries and technologies for soybeans produced in this area.

In this study, we determined the enterprises which occupy the special position in the network, using two types of "centrality" index was calculated, measuring the degree of influence on the development and maintenance of network. This measure clarifies the role of each company in its structurally unique position. In terms of the network centrality index, the position of Hokuren (Hokkaido agricultural cooperatives) stands out. The wholesalers and retailers closely associated with Hokuren show high values for both betweenness and the Bonacich centrality. This type of networking structure has, in a way, been systematically guaranteed by orderly marketing backed by the subsidy system for soybeans.

Keywords Soybeans • Food industry cluster • Network analysis • Centrality

1 About Soybean

Unlike other legumes, soybeans are low in carbohydrates and high in proteins and fats. Because of these properties, they have been an important source of nutrients in Japan since ancient times, complementing the nutritional value of grains such as rice and wheat, which are high in carbohydrates. The importance of soybeans

---

T. Morishima (✉)
Farm Management Division, National Agriculture and Food Research Organization (NARO), Agricultural Research Center, Tsukuba, Japan
e-mail: morisima@affrc.go.jp

extends beyond Japan. Soybeans are grown in several countries worldwide, where they are the main source of plant oils and plant proteins. The USA is the largest producer of soybeans in the world, followed by Brazil, Argentina, and China. These countries produce over ten million tons of soybeans per year. Other countries that produce over a million tons of soybeans per year include India, Canada, and Paraguay. Therefore, soybean production is particularly high in the Americas and in Asia. However, production in Europe is also increasing. Soybeans are a popular crop in all regions, except Africa.

Soybean production in Japan increased after the Second World War, but decreased by 1956 with the expansion of foreign currencies and import liberalization in 1961. Since 1993, however, crop rotation in rice paddy fields has been promoted as a means of production so that by 2009, the area under production recovered to 145,000 ha, giving a yield of 230,000 tons per year. Seventy percent of the demand for soybeans in Japan has been for processing as oil, and since the 1980s, this demand has remained stable at 3.5–4 million tons per year. But the supply and demand equations for soybeans have changed in recent years because of factors such as the US biomass and energy policy and Chinese demand increase. Nowadays, the demand for soybeans for oils in Japan has fallen to less than 2.5 million tons.

Most soybeans produced in Japan are used not for oil but directly as food. The demand for soybeans as food within Japan is approximately one million tons per year, and half of this demand is for tofu products; this demand has not considerably changed. Among fermented soybean products, the production and consumption of miso has been declining, but the use of whole soybeans in soy sauce, which is produced usually by defatted soymeal, is increasing; therefore, the total use is stable at 160,000–200,000 tons per year. The increased consumption of natto and soy milk has driven the growth in processed soybean products, and the demand for soybeans for food in general has gradually increased. The ratio of soybean self-sufficiency in Japan was 6.9 % overall and 20 % for the use of soybeans as food.

The self-sufficiency ratio for soybean production in Japan is low, as is the case with wheat, another important crop. This means that these crops cannot be relied upon in an emergency. As described above, the demand for soybean oil has been dropping recently, and one of the factors might have been a sharp rise in the price of imported soybeans since the autumn of 2003. Subsequently, the price returned to approximately 30,000 yen per ton, but there was another steep price rise in the autumn of 2006, when prices at one point were as high as 80,000 yen per ton. During this period, soybean oil was replaced by other plant oils such as palm and canola. The fluctuations in price of domestic soybeans have a much greater impact on domestic food manufacturers. Opening bids for soybeans produced in Hokkaido (named Toyomasari), which are generally close to the norm, ranged from 70,000 to 330,000 yen per ton (Fig. 7.1).

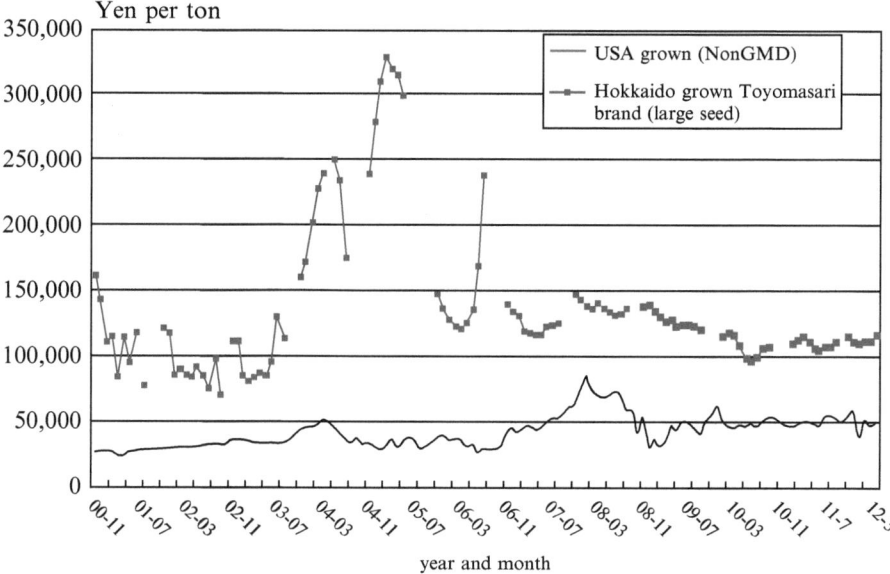

Fig. 7.1 Variation in prices of Hokkaido-grown and imported soybeans (Source: Japan Speciality Agriculture Products Association and The Tokyo Grain Exchange)

2 Soybean-Related Industries in Hokkaido

The reasons for price fluctuations in domestically produced soybeans are mismatches in the balance of supply and demand, particularly owing to yield instability. In 2002, prices of domestic products were approximately twice those of imported products. However, the harvest was poor during 2003–2004 because of extreme climatic events, and prices markedly increased. Because domestic production is unreliable, several producers are reducing their use of domestic products. The area responsible for the most soybean production in Japan is Hokkaido, which accounts for one fifth of all soybean production. Other significant production centers are Fukuoka and Saga prefectures in Northern Kyushu and several areas in the Tohoku region.

Hokkaido has the highest level of soybean production, but the lowest consumption of processed soy products in the country. Detailed consumption comparisons show that tofu, fried tofu, miso, and soy sauce consumption is below the national average. Only the consumption of natto is higher, but this difference has become smaller. Thus, Hokkaido has recently become regionally 100 % self-sufficient with regard to soybeans grown for food. However, "regionally self-sufficient" only implies that a region can produce enough to supply itself. In today's commercial world of large industries, raw materials are expected to be exported to large production

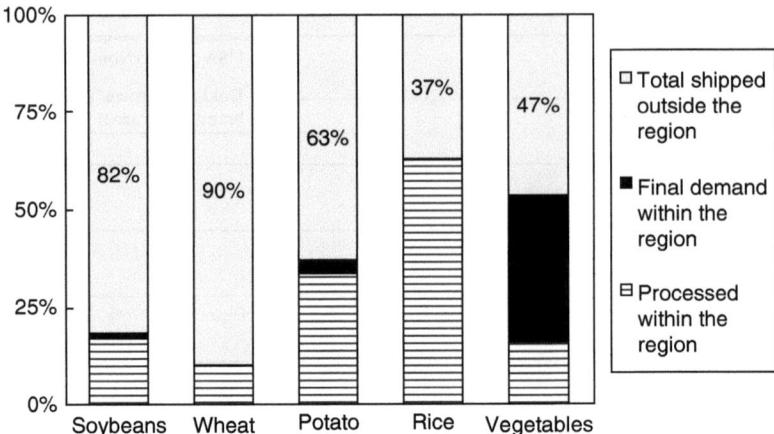

Fig. 7.2 Breakdown in demand for agricultural products grown in Hokkaido (Source: 2005 Input-output Tables for Hokkaido Region)

centers outside the region. Hokkaido has developed into a supplier of raw agricultural products to be exported to other regions. This is true not only for soybeans but also for other crops. It can be observed from 2005 regional inter-industry tables that 82 % of soybean production by value was exported from Hokkaido as unprocessed product (Fig. 7.2). Thus boosting the regional economy by adding value to Hokkaido agricultural products by processing will be a challenge.

With this goal, trials are under way in parts of Hokkaido to promote the formation and development of "food industry clusters" as part of a policy for promoting synergies between new products and new businesses, and cooperation between primary industries and food producers. When such industry clusters are proposed in Japan, the first step is for representatives from industry, academia, and the government to establish a core support structure called a "platform" to create an industry network around this structure. As this network expands, industries come to the area and clusters are formed. In Hokkaido, there are close to 30 such organizations in each regional city and other centers of population, where local businesses from various sectors congregate, exploring ways of using existing resources to their advantage. In the case of Hokkaido, the existing resources are often centered on agriculture, fishery, as well as forestry industries and food production. However in much of Hokkaido, there is little accumulation of industry and knowledge; therefore, their development has been stunted. In Hokkaido, only the Sapporo area has a well-developed infrastructure for industry and for producing unprocessed ingredients. A food production cluster is being developed for this area based on wheat and soybean processing. In terms of food production, the Sapporo area accounts for approximately two thirds of processed soybean products from Hokkaido. Tofu and semi-dry fried tofu account for approximately half of these, natto for 40 %, and soy sauce and miso for 90 %.

The "Sapporo Soybean Cluster" was formed in 2000 from a partnership between the 15 soybean processing companies, government, and academia in the area. The aim was to develop new industries and technologies for soybeans produced in the Sapporo area. This project was initially funded by subsidies, and at that time, its main success was in developing the new technology of producing fine soybean flour, which is useful for a variety of products. This program supported by a grant ended in the fiscal year of 2004, and the administrative function of this platform transferred from the regional government to a wholesaler. Subsequently, the membership has expanded, and includes cooked soybeans, wholesalers, trading companies, and agricultural cooperatives. As a result, several new processed food products such as tofu, fried tofu, natto, and roasted soybeans have been commercialized. Previously, many of these products failed to reach consumers even if they advanced to the prototype stage (see Morishima 2012).

3 Soybean Cluster Networks in the Sapporo Area

To paraphrase M.E. Porter's definition, a cluster is "a geographically proximate group of interconnected companies and associated institutions in a particular field," (Porter 1998, p. 199). Three important properties of a cluster can be gained from this definition: (1) specific industry, (2) interconnectedness, and (3) geographical concentration. Porter also has this to say about interconnections (point 2): "A cluster is a form of network that occurs within a geographic location," (ibid., p. 226). There are already soybean producers, processors, and distributors (1: specific industry) in the Sapporo area (3: geographical concentration). Forming a cluster requires a third condition, i.e., network formation established by cooperation between industry, academia, and government (2: interconnectedness). In the study described below, we aim to identify trends in networking by analyzing the structure for developing cluster networks and clarifying the role of different members.

Many of the corporations that became members of the "Sapporo Soybean Cluster", described above, are food processors. However, these food processors rely on distributors, such as wholesalers and retailers. In this study, we regard the members and associated client companies as a cluster and analyze the network between them. These data were gathered from the corporate information database of Tokyo Shoko Research, Ltd. (TSR). Of the 36 member corporations in 2006, 28 had registered in this database. The database collates commercial information, including the locations, capital, and sales volumes and their customers. For enterprises not listed in the database, information was gained from other sources such as chambers of commerce and public company records.

Various formal methods for modeling the structure of social and economic relationships between individuals and corporations have been developed. In this study, we first determined the enterprises which occupy the special position in the network, using the hierarchical clustering method applied to the overlaps of N-cliques. Subsequently, two types of "centrality" index were calculated, measuring the degree of influence on the development and maintenance of the network. This measure clarifies the role of each company in its structurally unique position.

3.1 The N-Clique

Sets of three or more nodes that are fully connected to each other in a network are called a clique. Cliques show close relationships but tend to be considerably small. This study accordingly identified N-cliques, showing relationships to reach each other up to the nth step. This set was evaluated using the overlap structure. If the N value is more than 3, the size becomes extremely large; therefore, this study targeted 2-step relationships. The number of 2-cliques for this cluster was 2,973 at the minimum size (3), and 1 at the maximum size (64). If the clique size was too small, it became an issue of missing the larger picture; if clique size was too big, the numbers were reduced so that only special cases were targeted. So the cliques that have intermediate size (15) or more were chosen. The evaluation of overlap structure for the group of 15 companies or more revealed a total of 372 cliques.

Among these cliques, 353 were revealed to be CO-OP Sapporo, all but one representing the Hokuren (Hokkaido agricultural cooperatives). These two companies appeared in the same cliques 339 times. The overlaps were grouped by hierarchical clustering, revealing a branching structure divided into five large enterprise groups (Fig. 7.3). Among these branches, those with the largest number of groups were Hokuren and CO-OP at its apex. The other enterprises at the upper levels of

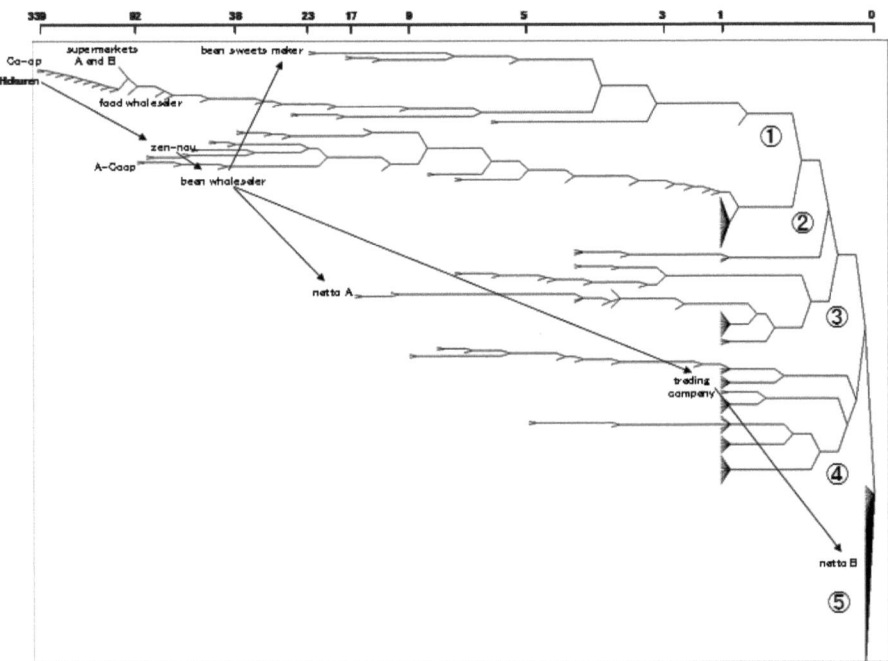

Fig. 7.3 Results from hierarchical clustering of 2-clique overlap structure. Note: The scale shows the number of overlaps

this branch were several soybean distributors within Hokkaido (supermarkets A, B and food wholesalers). In addition, these were connected to another block at a high level, which included a subsidiary retail company of Hokuren (A-COOP), a wholesaler from the head office of the cluster platform, and the National Federation of Agricultural Cooperative Associations (Zen-nou). These formed one large branch, which could be described as a single lineage called "*keitou*", forming the main cluster. Compared with these organizations, cluster member companies such as natto manufacturer A are small-scale.

3.2 Centrality

When the above points are expressed in terms of the network centrality index, whether with regard to enterprises doing business with enterprises with many business partners (the Bonacich centrality) or measured by the degree to which an enterprise mediates business relationships (the betweenness), the position of Hokuren stands out (Fig. 7.4). The wholesalers and retailers (CO-OP, supermarket A and B, and food wholesaler) closely associated with Hokuren show high values for both betweenness and the Bonacich centrality. In contrast, natto manufacturer A and a trading company only have a high betweenness score. These two enterprises have few transactions and form a group on their own, together with the specialized companies with which they interact.

When the cluster platform was established, manufacturing bean sweets was in the trial stage of product development, and since a lot of soybeans had not been

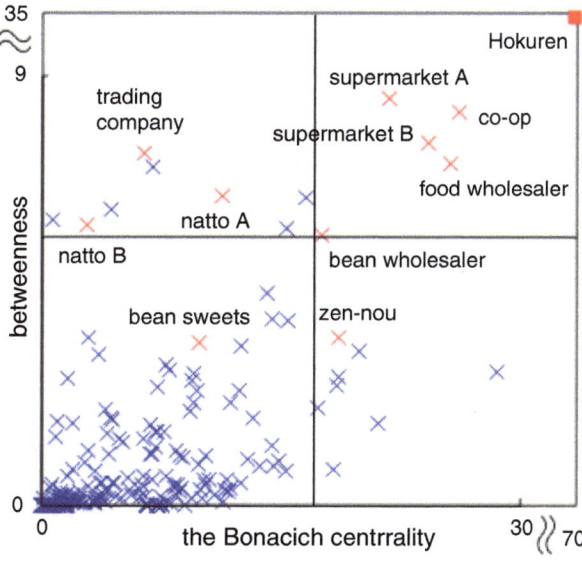

Fig. 7.4 Position of primary enterprises in terms of two centrality indices. Note: Axes indicate the upper 5 % enterprises

clearly organized, they could not use the soybeans targeted by their grant money; hence, black soybeans were used. Consequently, a soybean wholesaler became an active participant in the platform, and they could use both types of Hokkaido-grown soybeans. The distribution channels were developed, comprising Hokuren, Zennou, the soybean wholesaler and natto manufacturer A, and the channel comprising the soybean wholesaler, the trading company, and natto manufacturer B (Fig. 7.4). These operations developed natto products under the brand "Sapporo soybeans". As a result, networks expanded in the processed soybean food producer cluster, and soybean wholesalers played an important role. Thus, enterprises, including natto manufacturer A and the trading company, which bridge "structural holes" (Burt 1992) in networks, should be identified and included in the cluster to create new business opportunities.

4 The Challenge

The above analysis shows clearly that Hokuren is in a unique position of not being pushed out by competitors, and its role as a networking hub is clear. This type of networking structure has, in a way, been systematically guaranteed by orderly marketing backed by the subsidy system for soybeans. However, a management policy of "cross-sectional commodities" has been introduced, and subsidies solely for soybeans have ceased. Accompanying these changes, starting with the 2007 production season, the distribution system for soybeans has also markedly changed. In particular, because multiple distribution systems will not be required in future, producers can directly engage with buyers through contract farming. This distribution structure is not yet in place, but it is important to realize that such structural changes may take place.

There are two challenges for the future that new technology and new business opportunities may present. Currently, new rice products such as bread made from rice flour have been promoted throughout the country, and the development of technology using soy flour is being addressed in Hokkaido, not only in the Sapporo Soybean Cluster but also in parallel with other centers such as Asahikawa, Ebetsu, or Kitami. If a useful method is found in any of these places, it could lead to the production of Western-style sweets or pastries that so far have not been technically possible to make with soybeans. Bridging this technical gap would allow the creation of new business opportunities, making this an important possibility to explore.

References

Burt RS (1992) Structural holes: the social structure of competition. Harvard University Press, Cambridge, MA

Morishima T (2012) The network structure analysis of the food industry cluster. Norintokei Kyokai, Tokyo [in Japanese]

Porter ME (1998) On competition. Harvard Business School Press, Boston

Chapter 8
The Competitive Advantages of Green Tea Clusters in Japan

Yuko Akune

Abstract The purpose of this paper is to consider green tea clusters based on the concept of clusters and competitive advantage in Porter (On competition. Harvard Business School Press, Boston, 1998). After specifying the green tea cluster in Shizuoka, it considers activities within the cluster in terms of competitive advantage. As a result of the comparison of several green tea clusters, there were three main features. The first feature was the existence of technical advisors assisting producers within the cluster. They could provide strong professional advice to producers and proposed significant improvements for promoting products in Shizuoka and Kyoto. The second was the use of a trade fair for access to public goods. A fair specifically for green tea was held uniquely in Shizuoka. Participants were able to lower their transaction costs, including investigation time and costs, by being able to compare various commodities at the same time and place. The variety of products on display let producers observe the range of others' products and functioned as an incentive for their own product improvement. The last feature involves "demand conditions". The existence of plentiful demand from households in Shizuoka and strong pressure from special customers in Kyoto pushed producers to provide reasonable product prices and quantities and to increase quality. Hence, the clusters that had technical advisors associated with them had competitive advantage in the green tea clusters in Japan. These included leading the quality frontier, the ease of access to public goods to encourage the incentives for producers and enhancing demand conditions.

Keywords Green tea cluster • Competitive advantage • Technical advisors • Access to public goods • Demand conditions

---

Y. Akune (✉)
Faculty of Economics and Business Administration, Reitaku University, Kashiwa, Japan
e-mail: yakune@reitaku-u.ac.jp

1 Introduction

The purpose of this paper is to consider the green tea clusters based on the concept of clusters and competitive advantage in Porter (1998). Japan is facing the issue of a declining population, which has damaged rural regional economies in particular. Since the late 1960s, domestic rural economies have developed as a result of moving the locations of manufacturing firms and their plants from urban areas in Tokyo, Yokohama and Osaka. However, since the late 1980s, many manufacturers have demanded lower labor costs and larger market share, and as a result, rural regions have been endeavoring to maintain their economies with their own resources.

The main resource in each rural region is agriculture, producing agricultural products based on the capacity of the natural conditions. It is possible to encourage rural economies if value, such as food processing, is added to agricultural products whilst they are still within the region. The Japanese government, especially the Ministry of Agriculture, Forestry and Fisheries, has pursued various policies for this purpose, such as a scheme of clusters of food industries called the "shokuryo-sangyo cluster", a program of agri-food industries called the "rokuji-sangyoka", which means activities collaborated on by agriculture, manufacturers and service industries, and a joint program regarding collaborative links between agriculture and related industries with the Ministry of Economy, Trade and Industry called the "no-sho-ko-renkei". A lot of projects have been accepted and supported under these policies. Each project has the potential to encourage rural economies, if it succeeds. One of the factors for this success consists of the competitiveness of each activity. This importance of competitive advantage in a cluster was identified by Porter (1998).

It is necessary to consider the competitive advantage of collaborative activities between agriculture and related industries as the source of rural economic growth. Examples include previous studies on clusters related to agriculture and food. There are also many case studies, such as a wine cluster in Yamanashi (Kageyama et al. 2006), a plum cluster in Wakayama (Saito 2007), a green tea cluster in Shizuoka (Akune 2009, 2011), and soybeans in Hokkaido (Morishima 2012). Kageyama et al. (2006) presented an agglomeration of the wine industry and linking between wine producers within the cluster. Saito (2007) indicated the existence of a diamond cluster, suggested by Porter (1998), in the plum cluster. Morishima (2012) denoted the importance of networks within clusters related to soybeans by using a method of network analysis. Moreover, Akune (2009; 2011) described the existence of a cluster for a food system of green tea in Shizuoka, as in other previous studies. However, they were not sufficient for considering the competitive advantage of a cluster. Agricultural activities are restricted by natural conditions, although these conditions are not absolute constraints. This results in the fact that the same agricultural products are grown in various places. Green tea is also produced in many prefectures: for instance, Shizuoka, Kagoshima and Kyoto, and so forth. On the other hand, there are obvious differences in the circumstances of production among prefectures. This chapter focuses on the competitive advantage in the green tea clusters.

This paper consists of four sections. Section 2 describes the state of green tea production and distribution in the green tea food system in Japan. Section 3 identifies the green tea cluster in Shizuoka. Section 4 considers competitive advantages in the green tea clusters by describing activities and networks of information in the green tea cluster in Shizuoka and compares this with the green tea clusters of other prefectures. Finally, the Conclusion summarizes and comments on the findings of this paper.

2 The Food System for Green Tea in Japan

This section includes an overview regarding the upstream and downstream production of the green tea industries in Japan. There are two producers that supply this industry: tea farmers upstream and tea blenders downstream. First, the tea farmers cultivate tea leaves as a raw material, and then produce semi-finished green tea as an intermediate called "Ara-cha" by steaming and drying fresh tea leaves. Some intermediates are made by tea farmers' cooperatives but the farmers still own them. Tea leaves are not transportable outside of native growing areas in order to maintain freshness, while semi-finished green tea is available for inventory and trade outside native areas without restrictions. On the other hand, the tea blenders, who are the downstream producers, generate finished green tea called "Shiage-cha" by re-roasting and blending several intermediates.

To meet the demand for the product, the finished green tea is distributed to the following three final destinations: households, beverage manufacturers, and firms making confectionery, bread and dairy products. Households consume 73 % of finished green tea, which is typically brewed in a teapot. Almost 25 % is used as intermediate inputs for producing bottled tea and the remaining 2 % is used for green tea flavoring and for coloring confectionery, bread and dairy products.

There are mainly two properties of green tea production in Japan. One feature is the high concentration in Shizuoka Prefecture. Table 8.1 shows the output values for green tea processing in 2010. The total of domestic production was 363 billion yen in the table, in which the highest ranking prefecture was Shizuoka, at 182 billion yen,

Table 8.1 Outputs in the food system of green tea in 2010. Unit: billion JPY

	Tea leaves	Semi-finished green tea	Finished green tea	Total
Japan	74	64	225	363
Prefecture				
Shizuoka	30	24	128	182
Kagoshima	17	23	10	51
Kyoto	4	2	32	38
Fukuoka	3	2	9	14
Mie	6	4	4	13

Sources: Ministry of Agriculture, Forestry and Fisheries "Statistics of Production and Agricultural Income in 2010", and Ministry of Economy, Trade and Industry "Census of Manufactures in 2010"

and Kagoshima Prefecture was second, at 51 billion yen. That means that Shizuoka yields half of the domestic green tea production and 3.6 times that of the second highest prefecture. For this situation, a geographic concentration index, suggested by Ellison and Glaeser (1997), regarding green tea manufacture that consists of plants of semi-finished and finished green tea had a higher ranking in Japanese manufacturers, as in Akune and Tokunaga (2003). Another point is the proportions of each production stage, for which there are two patterns. The average proportion for output of tea leaves is 20 %, for semi-finished green tea 18 %, and finished green tea 62 %. Thus, the production of finished green tea earns the highest yield in the food system for green tea. This situation is the first pattern and is shown in Shizuoka, Kyoto and Fukuoka Prefectures. On the other hand, the proportion of the upstream growing of tea leaves and producing of semi-finished green tea is higher than that for the production downstream. Kagoshima is a typical prefecture: the proportion of the output of tea leaves is 34 %, of semi-finished green tea 46 %, and finished green tea 70 %. In Mie, although the order of tea leaves and semi-finished green tea is opposite in comparison with Kagoshima, the sum of these proportions is higher than for finished green tea. These differences are a reason why semi-finished tea has been traded among prefectures. As a result, Shizuoka and Kyoto Prefectures have a high demand for green tea materials, while Kagoshima and Mie Prefectures are the main supply regions.

Still dealing with the demand-side of the industry, Table 8.2 shows the annual household expenditure on green tea in cities that are situated in high production prefectures in 2010, in which the ranking denotes the order for all the surveyed cities. The term "expenditure on green tea" in this table denotes the expenditure on green tea that is brewed in a teapot, but this excludes bottled tea. The top four cities, Hamamatsu, Shizuoka, Kagoshima, and Kitakyushu, are all located in green tea producing prefectures. Also, the percentage of expenditure on green tea for beverages in the two cities in Shizuoka, at approximately 20 %, stands out from the other cities. Cities in Kagoshima and Fukuoka Prefectures spend almost 15 % on green tea for beverages and are higher than the domestic aver-

Table 8.2 Annual expenditure on green tea in cities in 2010. Unit: yen/household

		Annual expenditure		
		Rank	Green tea	Ratio of green tea to beverage
Japan		–	4,424	9 %
City	Prefecture			
Hamamatsu	Shizuoka	1	9,860	20 %
Shizuoka	Shizuoka	2	8,821	21 %
Kagoshima	Kagoshima	3	6,590	14 %
Kitakyusyu	Fukuoka	4	6,405	15 %
Fukuoka	Fukuoka	23	4,285	12 %
Kyoto	Kyoto	25	4,148	9 %
Tsu	Mie	29	3,884	9 %

Sources: Ministry of Internal Affairs and Communication, "Survey of Household Economy in 2010"

age of 19 %. Although the annual expenditure on green tea in Kyoto City and Mie City is below the average expenditure, the percentage is still at the average value. It is clear that households have a strong tendency to spend on green tea in regions that produce the product.

3 Identification of Participators in the Green Tea Cluster in Shizuoka Prefecture

According to Porter (1998), a cluster is defined as "*a geographically proximate group of interconnected companies and associated institutions in a particular field*" (p. 199). This section identifies the Green Tea Cluster (GTC) in Shizuoka, based on Porter's paradigm of the cluster.

It is important to describe the food system of green tea on a geographical basis for identification of the participants in the CTC in Shizuoka. Table 8.3 shows this system by region in Shizuoka according to the type of green tea product: tea leaves, semi-finished green tea, and green tea manufacture mixed establishments of both semi-finished and finished tea. For the upstream for green tea production including tea leaves and production of semi-finished tea, the ratio of outputs of tea leaves and semi-finished tea, which have a relationship to tea farmers, was 33.6 % in Makinohara in the prefecture. Also, 26.5 % of green tea manufacturers, including both semi-finished producers and tea blenders downstream, were located in the Makinohara area. Hence, all production stages are wholly concentrated in the Makinohara region to the west.

For the downstream of green tea production, there were almost 284 established tea blenders in Shizuoka in 2010, according to the Census of Manufacturers by METI, which operated 20 large tea blending firms that generated one billion yen in annual sales; the remaining companies were small and medium-sized tea blenders. Large tea blender firms produce 50 % of the finished green tea made in Shizuoka. They sell their products to commercial-scale utility customers, such as beverage manufacturers, food processing manufacturers, and large retailers. On the other hand, 42 % of green tea is mainly sold by small and medium-sized tea blenders to specialty tea shops. The scale of sales between tea blenders and their customers is positively correlated.

There are connectors from the upstream to the downstream in the green tea system in Shizuoka: these include tea cooperatives, tea wholesalers called "Saitori", and the Shizuoka Tea Market. Forty-five percent of traded semi-finished green tea goes through tea cooperatives, while 44 % is handled by non-tea cooperatives such as tea wholesalers and the Shizuoka Tea Market. This production and distribution is a vertical chain in the GTC.

Considering the cluster as a whole, it is necessary to look at a horizontal chain that exists in this food system. According to Porter (1998), a "horizontal chain"

Table 8.3 Green tea production in Shizuoka

Region	Area of tea farm (ha)	Ratio	Production of semi-finished tea (t)	Ratio	Outputs of tea leave and semi-finished tea (10 million yen)	Ratio	Location of green tea manufacture	Ratio
Makinohara	6,010	29.2 %	13,088	35.7 %	2,223	33.6 %	288	26.5 %
Tyuen	3,988	19.4 %	8,054	22.0 %	1,385	20.9 %	171	2.6 %
Shida	2,682	13.0 %	5,022	13.7 %	854	12.9 %	138	2.1 %
Shizuoka	2,100	10.2 %	3,240	8.8 %	717	10.8 %	159	2.4 %
Fuji/Nansun	1,867	9.1 %	2,716	7.4 %	435	6.6 %	142	2.1 %
Tenryu/Mori	1,349	6.5 %	1,215	3.3 %	319	4.8 %	64	1.0 %
Kawane	1,043	5.1 %	1,288	3.5 %	330	5.0 %	60	0.9 %
Seian	948	4.6 %	1,419	3.9 %	257	3.9 %	5	0.1 %
Seien	380	1.8 %	494	1.3 %	69	1.0 %	29	0.4 %
Hokusun	140	0.7 %	71	0.2 %	8	0.1 %	10	0.2 %
Izu	93	0.5 %	30	0.1 %	16	0.2 %	19	0.3 %
Shizuoka Pref.	20,600	100.0 %	36,637	100.0 %	6,613	100.0 %	1,085	16.4 %

Sources: Shizuoka Prefectural Government "Report on the Green Tea Industry in Shizuoka 2003", NTT "i-Town Page"

Note: "Location of Green Tea Manufacture" includes both semi-finished tea producers and finished tea producers. I selected the category, "Seicha-gyo" from "i-Town Page" published by NTT

involves industries producing complementary products and services. The processes of growing and making green tea have been mechanized since the late 1960s. Almost all processes in green tea industries are presently mechanized and automated. Hence, this has a large effect on productive efficiency and quality of products. Thirty-four percent of green tea machinery firms are also located in the Makinohara area. These industries are recognized as a *"geographically proximate group"* and a horizontal chain of green tea production in Shizuoka. Therefore, tea machinery firms are participants of the GTC in Shizuoka.

The cluster members are included not only as producers and distributors but also as coordinators. Coordination is a particularly essential factor that ensures the maintenance of the local brand "Shizuoka cha (green tea made in Shizuoka)"; this process involves a lot of interested parties: tea farmers, distributors, and tea blenders. Three institutions, the Chamber of Tea Association of Shizuoka Prefecture (CTASP), the JA Shizuoka Keizairen (JASK), and the Tea Commerce and Industry's Association of Shizuoka Prefecture (TCIASP), accommodate and coordinate the requirements from producers and related firms in the industry. The Nihon Seicha Kikai Kogyokai (NSKK) is an association of green tea machinery companies and a member of TCIASP. In addition, the Shizuoka Prefectural government has overseen and subsidized CTASP and monitored green tea producers.

There are four institutions supporting and leading innovation for more productive efficiency and for higher quality in the GTC in Shizuoka Prefecture: the Shizuoka Prefectural Research Institute of Agriculture and Forestry (SPRIAF), the Kanaya Tea Research Station of the National Institute of Vegetable and Tea Science within the National Agriculture and Food Research Organization (KTRS), Shizuoka University, and Shizuoka Prefectural University. SPRIAF and KTRS are located in the Makinohara area. SPRIAF especially is a member of the production committee of CTASP: this institute researches and develops technologies and new products for green tea production from growing tea leaves to making finished green tea. In the development of new machinery and equipment, it has collaborated with several tea machinery firms since the 1960s. Also, researchers and advisors at SPRIAF instruct producers on how to use new technologies and provide related knowledge at various lecture classes and training workshops.

Finally, requirements and complaints from the upstream and downstream sometimes conflict. This leads to accommodations necessary to maintain the standard quality of green tea produced in Shizuoka and to encourage better production practices, while solving problems dealing with production and distribution. The Tea and Agricultural Production Division at the Shizuoka Prefectural Government (TAPD) is a government agency for green tea in Shizuoka Prefecture. TAPD has overseen and subsidized CTASP and monitored green tea producers in several prefectural ordinances for CTASP and the green tea industry. It has internationally collected and published various information regarding tea production, drinking style, and health effects. Also, it encourages firms and studies that contribute to the innovation of the green tea production field and consumption field.

4 The Competitive Advantages of the Green Tea Cluster in Shizuoka Prefecture

A cluster concerns not only the group that constitutes firms and industries in a particular area, but also affects the competition with the cluster. Porter (1998) emphasized the following three ways to affect competition: (a) increasing the productivity of constituent firms and industries, (b) increasing their capacity for innovation and productivity growth, and (c) stimulating new business information that supports innovation, and expands the cluster (p. 213).

This study focuses on point (a). Green tea is a traditional beverage which has been consumed for thousands of years. Current technologies and skills are based on traditional knowledge and styles. Although most processes are mechanized, they are merely a substitution for the producer's hand with machinery and can be used anywhere. Thus, it is not possible to expect anything in terms of points (b) and (c) within the existing GTC. On the other hand, there is obviously the existence of differences within regional competition as described in the section above. It is supposed that the clusters which have competitiveness have competitive advantages. It is necessary to observe the actual conditions of the cluster and place particular focus on the activities to increase the productivity of producers within the GTC of Shizuoka.

Porter (1998) indicated the five points that describe what gives clusters competitive advantages: (1) access to specialized inputs and employees; (2) access to information; (3) complementarities; (4) access to institutions and public goods, and; (5) incentives and performance measurement.

The ease of access of points (1) and (2) contributes greatly to a decrease in transaction costs. The quality of green tea made solely with tea leaves depends on the material's quality. The essential element in lowering transaction costs is the information about the buyer's requirements. There are 221 establishments producing semi-finished green tea, and 284 establishments producing finished green tea. Generally, tea blenders change 30–40 % of their intermediate producers, depending on the demand for tea, and are required to source new partners in a limited period of time. This means that the highest peak of cultivation of tea leaves and contracting between upstream and downstream producers is 30 days during April and May. They make a contract regarding annual amounts during this period; both types of producers, however, have limited time to search for new partners as they are concentrating on the production of their products. These producers are connected through tea cooperatives and tea wholesalers, and play a vital role in coordinating the fields of tea-leaf growing and green tea processing. Tea cooperatives and tea wholesalers not only make semi-finished tea, but also provide information regarding quality and quantity. They function as a link between semi-finished tea producers and finished tea producers, and distribute commodities and related information to both producers. Tea wholesalers generally provide three to five samples of the semi-finished green tea together with information regarding tea-leaf production in farms to tea blenders. Each tea blender negotiates and contracts with each tea wholesaler and tea cooperative in their own firm or factory. According to an interview investigation at

CTASP, they can exchange various pieces of information regarding the requirements of tea blenders, as well as the possibility of supply from the tea farmers. They make arrangements, and sometimes rearrange them based on their own information and any new information that they may receive. They also provide tea farmers with feedback concerning the needs of producers in downstream industry. Tea-farmers in upstream industry, and blenders in downstream industry can immediately gain information regarding their own partner's situation and demand through tea wholesalers and tea cooperatives. This information is beneficial not only in the short term, but also in the long term as it aids them in making production and investment goals.

Horizontal information is also important for the competitive advantages of clusters, and exists in two paths in the GTC of Shizuoka. One path is a technological link by tea machinery firms that provide various types of machinery and equipment to farms and green tea processing plants. From the interview investigation, it was evident that it is possible for them to acquire their customer's requirements and circumstances, because they are available to maintain and instruct on their machinery 24 h per day during the highest peak period mentioned above. During this time, through discussions with the engineers, farmers and tea blenders involuntarily source various pieces of industry information, such as the status of other factories. Therefore, they are also important in collecting and spreading horizontal information. The second path involves producers who belong to the same industry. It is natural to develop a horizontal network of producers when they live in the same community. In addition, the opportunity to meet with fellow producers outside of the community within the prefecture is provided by associations such as JASK, TCIASP, and CTASP. They hold various lectures and workshops with the aim of improving the skills in the prefecture. Participants totaled over 1,300 people, and they were provided with new knowledge and skills at workshops in fiscal year 2002. At the same time, they are able to form relationships and share their circumstances. Hence, horizontal information provides opportunities for members of the same profession to share issues and discover solutions.

The requirements and claims between tea farmers and tea blenders sometimes conflict. It is necessary to consider them as a whole green-tea industry in order to maintain their own brands and improve competitive advantages. The requirements of the tea farmers are provided to JASK, while the needs of the tea blenders are provided to TCIASP. Mutual requirements are discussed and adjusted at CTASP, which consists of the following three committees: tea production and distribution, customer promotion, and finance. All committees are comprised of JASK and TCIASP. The Shizuoka Prefectural government, SPRIAF, the Shizuoka Tea Market, and members of the prefectural assembly join the committees when necessarily. The considerations for tea production and distribution in fiscal year 2002 were achieved primarily through a productive and cooperative link between upstream and downstream green tea production through productivity improvement, infrastructure development and mechanization on farms, promotion of recommended tea cultivars, an operational system for sanitation and traceability, and the provision for competition in the national tea market. The considerations for tea promotion in fiscal year 2002 were the advertising of Shizuoka green tea, fostering of green tea consumption, accommodation of a common element for their products, and organizing big events. The finance committee plays a role in monitoring the plans and acts of other committees, thus, information from tea farmers and tea blenders is officially assembled and conciliated at CTASP.

The results produced by CTASP require feedback to participants in order to benefit their validity. Information regarding the activities of CTASP is announced at the annual general meeting, and also printed in CTASP's bulletins which are published biannually. Additionally, the local newspaper, the Shizuoka Shinbun, and the local pages of other newspapers produce a daily column exclusively about the green tea industry and its related events. As a result of these measures, information from CTASP is shared throughout the members of the GTC within Shizuoka Prefecture and is available by the evening of the following day, at the latest. This information is not only shared, but also formally and informally accumulated by individual producers, firms and several industrial associations, and local institutions. Furthermore, the amassed information becomes experience and knowledge in the GTC, which provides the resources for the competitive advantages of clusters.

Complementarities in (3), which are indicated as holding a competitive advantage by Porter, involve shareable values for the participants of clusters. In the case of the GTC in Shizuoka, this includes having some effective marketing of the local brand and considering the customer's taste in green tea, maintaining and improving the regional brand, encouraging the consumption of green tea in households, and establishing common criteria and manuals and having these monitored.

The producers downstream in clusters need to take into account the taste trends of customers. Even though each firm should market their own products, it is difficult to explore the entire trends of consumption and possibilities for new products, such as non-food products, due to the cost-and-benefit issue. In the GTC of Shizuoka, TCIASP is an organization of tea blenders who represent the green tea industry, and whose role it is to survey the situation of green tea distribution, consumption, and the customer's needs. According to literature on the history of the prefecture's green tea industry, the consumption of tea and related products have been surveyed since the late 1960s. Until the 1980s, the purpose of the survey was mainly fact-finding, and was targeted at both households and related industries, such as tea retailers. The purposes of the research began to change as of the 1990s. Firstly, their target moved from being business-related to the final consumers. Secondly, they started to cover a large range of purposes including branding of green tea and sample surveys of new products. Moreover, they searched for new non-food products that utilized green tea, such as toothpaste and cleansers made up of green tea components. Generally, it is noted that the general public recognize green tea as having health efficacy and the ability to decrease smell.

For the green tea industry, the sale of green tea leaves produces a higher profit than the sale of materials for green tea bottled beverages or sanitary products. For this reason, TCIASP is in charge of public relations, and encourages the brewing of green tea in a pot within the GTC. Their target is not only adults, but also students attending junior and elementary school. The purpose of the lecture is to demonstrate how to brew green tea to ensure the best taste. These activities have contributed to Shizuoka Prefecture having the largest domestic consumption of green tea in households in this region, as mentioned above. At the same time, to further increase the consumption of green tea, TCIASP focuses on the high efficacy of green tea for health. Studies on the health effects of green tea have been subsidized and recognized, and the results are reported at the debriefing meeting, which is then covered by the media to ensure the information is passed on to the customers.

The common standards and manuals work as complementarities for the participants to keep the standards of quality for their own brands. More recently, customers require strict guidelines with respect to food safety, meaning that there is an aim to reduce the environmental impact of growing tea leaves on farms, ensuring the cleanliness of the manufacturing process, and ensuring that no needless food additives are included. In Shizuoka, a manual of sanitation control for tea farms and semi-finished green tea factories was drawn up in 2004, as well as one for finished green-tea plants in 2005, and these were produced by CTASP and Shizuoka Prefectural government. In their support, the NKK, which is the association of tea machinery firms, selected and dispatched two firms from their members. To monitor the producers, the prefectural agency conducted a direct investigation into semi-finished green tea, and a sample survey of commercial products to check the quality of finished green tea. Thus, they have maintained their own criteria for the quality of the whole green tea industry in Shizuoka.

The ease of access to local institutions and public goods including quasi-public goods in (4) fosters improvement of an individual producer's skill, and as a result, contributes to a rise in competitive advantage. For a whole cluster, it contributes to a decrease in the cost of research and development (R&D) and worker training. SPRIAF has been in charge of the R&D institute in the GTC in Shizuoka. As mentioned above, current processes are mechanized. According to the history book on the green tea industry by CTASP, SPRIAF has been in joint research and development with every tea machinery company since the late 1950s. It has also provided more effective operation methods to decrease production costs and increase quality for producers within the GTC. In addition, the R&D areas of SPRIAF have expanded from processing on farms and in factories, to creating new products. Until the mid-1980s, the main focus was on convenient drinking styles, such as instant tea, tea bags, and brewing green tea in cool water. SPRIAF studied various types of green tea to extend the frontiers of green tea utilization, for example, as edible whole leaves, condensed tea, and as materials for other processed foods. These products have been developed in collaboration with SPRIAF and JASK, with a focus on the efficacy of the tea components for health. Private companies have also aided in practical examinations. Thus, local institutions play a role in providing the results of their own R&D to farmers and producers within the GTC.

Participants in the GTC have the opportunity to gain valuable insight and experience regarding proficiency in their particular skill. In particular, JASK has run various courses and training workshops. In fiscal year 2002 the themes of these included: systematic learning of necessary techniques and skills in tea production, institutes for the acquisition of licenses for high-quality improvement, training instructors for the techniques of semi-finished tea and improving farming management, and experiments in good practice for higher products. These events are conducted in accordance with the decisions of CTASP. In addition to this, JASK conducts its own workshops. The objectives of these are varied and include: techniques of tea production, factory operation and management, and the efficacy of green tea. All factories of semi-finished tea can access these workshops, regardless of whether or not participators are members of an agricultural cooperative. JASK staff visit and report to factories which do not participate in workshops. According to the interview research, although tea cooperatives treat 30 % of semi-finished tea within Shizuoka prefecture, the information that JASK holds is distributed to 70 % of semi-finished tea producers, thus providing opportunities to improve productivity which in turn has an effect on public goods throughout the cluster.

Moreover, due to improved production technology and product quality in Japan, a national green tea show has been held annually. In Shizuoka, teams of advisers are organized to enhance competitiveness in the national competition. They visit the farms and factories of competitors to coach them in how to win the show. A total of 2,311 producers took advice from them, out of which 100 candidates were selected for the 2002 competition. As a result, the green tea industry in Shizuoka has regularly turned out first place winners in the national competition. Additionally, there is a prefectural show as well as regional shows. As with the national competition, producers can take advice to improve their product quality for when they exhibit it.

The points described above relate to the effects of corporations. Incentive and performance measurement of producers in (5) encourage competition. As a result of individual competition, clusters gain a competitive advantage (Porter 1998). In the case of the GTC in Shizuoka, the shows give participants the impetus to produce green tea with a higher quality. In fiscal year 2002, apart from the national show, there was also one prefectural show and 47 regional shows. Winners were publicly announced in various media and in reports by related associations and the prefectural government. Producers were permitted to show this on their products. Competitions such as these work as an incentive to raise individual skill and foster an increase in the competitive advantage of the cluster.

Finally, the features of green tea clusters in Shizuoka are further clarified by making a distinction between Shizuoka and other prefectures. Figure 8.1 illustrates green tea clusters in Shizuoka, Kyoto and Mie Prefectures. There is little difference between the members of other clusters and those within the GTC in the following areas: producers upstream and downstream, related associations, prefectural institutes, and prefectural governments. However, there is a difference in the type of access to public goods among prefectures. First is technology, which plays a key role as technology advances, in assessment and advice for quality and quantity. The prefectural tea institute in Kyoto investigates everything from tea leaves to finished green tea in the same way as the institute in Shizuoka. Also, at the prefectural tea institute in Kyoto, there are experts who can point out detailed flaws in products and offer solutions. As a result, many producers in Kyoto and Shizuoka have won awards for both semi-finished tea and finished tea at the national competition show. On the other hand, the institute in Mie focuses on studies regarding production of tea leaves and semi-finished green tea. Also there is no adviser who is able to generate better quality finished green tea. Consequently, producers in Mie have mainly received awards for semi-finished tea.

Another difference is the trade fair. The Shizuoka Prefectural government has held a trade fair every year, promoting its green tea, open to distributors, retailers, processed food firms, and even consumers. Customers are available to directly

8 The Competitive Advantages of Green Tea Clusters in Japan

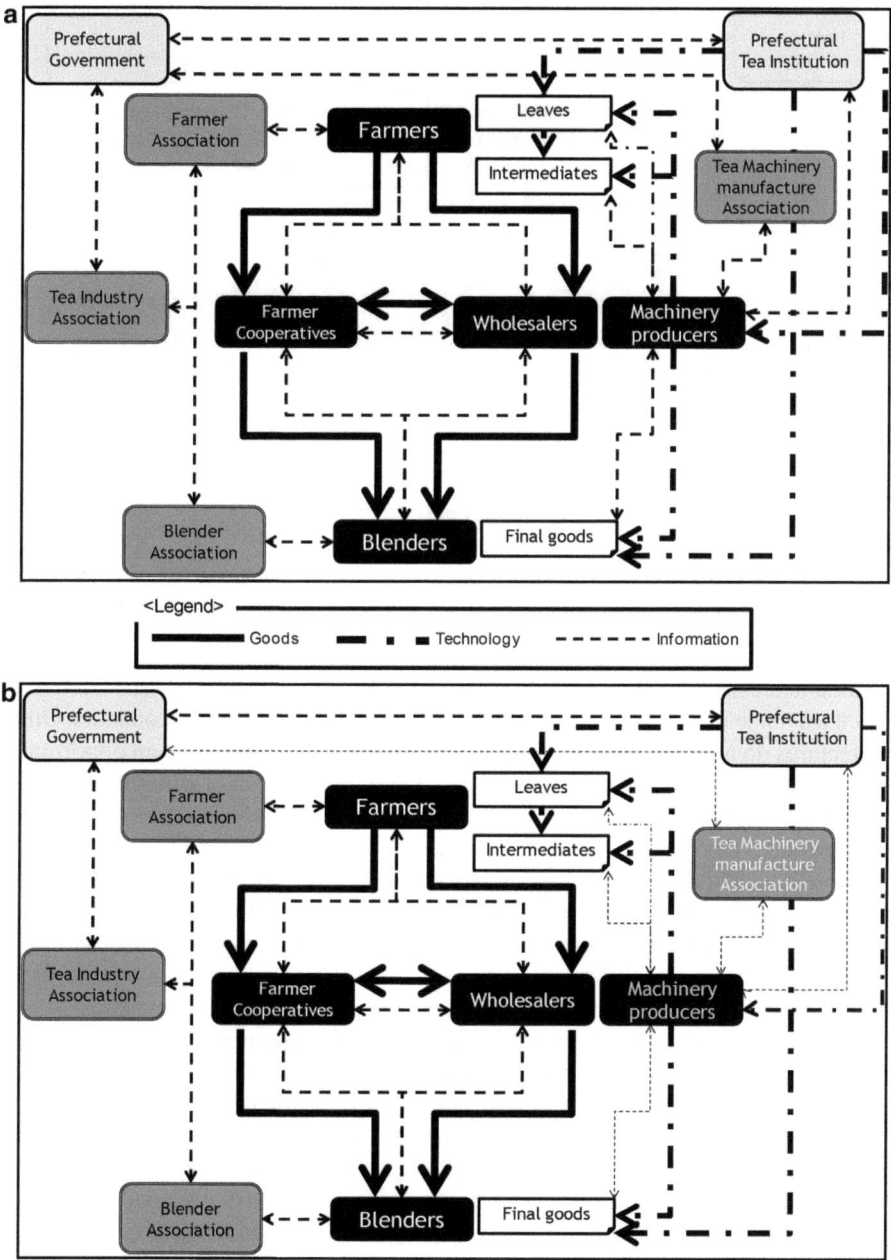

Fig. 8.1 The green tea clusters in Shizuoka, Kyoto and Mie prefectures. (**a**) Shizuoka Prefecture, (**b**) Kyoto Prefecture, (**c**) Mie Prefecture. Note: *Line thickness* shows size of quantity, strength of information and connection

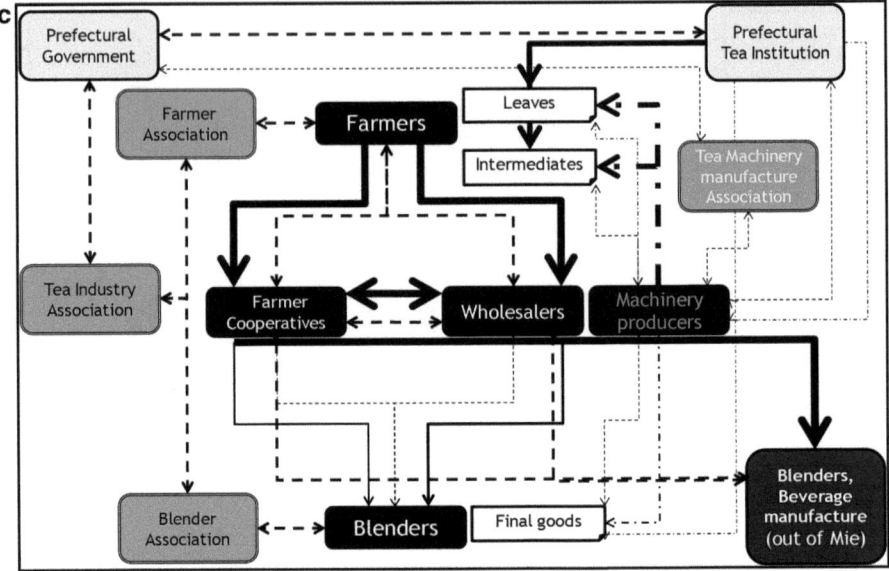

Fig. 8.1 (continued)

communicate with producers and, at the same time, compare various varieties of green tea made in Shizuoka. It is also possible to talk business with producers. Other prefectures do not hold tea trade fairs of this nature, nor do they have an opportunity to promote their own products at food fairs, which are combined with other agricultural products and processed food. The ability to have a trade fair can be seen as a public good for producers.

Moreover, Porter (1998) emphasized the importance of demand conditions as one of the sources of the competitive advantage of a cluster. From the data regarding average household expenditure on green tea by some cities in related prefectures in Sect. 2, they are at very different levels, even if the prefecture is in the top five areas of green tea production. The consumption of green tea in Hamamatsu City and Shizuoka City, which both belong to Shizuoka Prefecture, are the highest and second highest in not only expenditure, but also in the expenditure ratio of green tea to other beverages: approximately 20 %. This ranking is proportional to the output of finished green tea in downstream industry, but not tea leaves and semi-finished green tea upstream. Nevertheless, Kyoto Prefecture is an exception. In the interview survey conducted concerning a tea blender in Kyoto, the existence of notable tea families in a traditional Japanese tea ceremony is still very much active today. They have been ordering new and original green tea every New Year. Some tea blenders have purveyed the products that meet their strict requirements every year. Their products are delivered to their followers as a specially produced product by each tea family head. People unrelated to these families are able to obtain the tea in the blenders' shops and nationwide department stores. These circumstances have seen the quality of green tea improve and maintained not only for the tea blender's own brand, but also for the regional brand in Kyoto. At the same time, the tea blenders demand that the semi-finished tea be stable enough for their products. Thus, some of the special customers expect the highest degree of product quality from the producers in downstream industry. This then has vertical and horizontal effects on the improvement of competitive advantage within the cluster. In the case of green tea clusters, the demand conditions assume high expenditure by households and an extremely high level of input from experts.

5 Conclusion

The purpose of this paper was to consider green tea clusters based on the concept of clusters and competitive advantage in Porter (1998). The green tea cluster in Shizuoka, the prefecture where the most green tea is produced in Japan, was identified as consisting of tea farmers and tea blenders who are producers in upstream and downstream industries, tea cooperatives and tea wholesalers who connect and arrange commodities and information between producers, tea machinery firms as members of the horizontal chain supporting technologies and distributing information among participants in the GTC, the prefectural green tea institution which researches and develops new products and machinery, four associations, and the prefectural government. These formats were constructed in every prefecture that produces green tea: Kyoto and Mie. Kyoto especially had the same features as Shizuoka, which demands intermediates from other prefectures, although they produce tea leaves and intermediates. On the other hand, Mie lacked several functions within its own cluster.

As a result of the comparison of several green tea clusters, there were three main features. The first feature was the existence of technical advisors assisting producers within the cluster. They could provide strong professional advice to producers and proposed significant improvements to lead products. These advisors were found in the clusters in Shizuoka and Kyoto Prefectures, which have turned out frequent winners in a national competition every year. The second was the use of a trade fair for access to public goods. A trade fair provided ease of connection between producers and customers at the same time. A fair specifically for green tea was held only in Shizuoka. Participants were able to decrease their transaction costs including investigation time and costs by being able to compare various commodities at the same time and place. The variety of products on display let producers observe the variety of others' products and functioned as an incentive for their own product improvement. The last difference involved "demand conditions". Households in Shizuoka consumed the most green tea in Japan. However, although Kyoto's household expenditure ratio of green tea was at the national average level, it had the tea experts, who are notable tea families in the traditional Japanese tea ceremony, almost all of them regularly active in Kyoto. The existence of plentiful demand from households and strong pressure from special customers pushed producers to provide reasonable product prices and quantities and to increase quality. Hence, the clusters that had technical advisors associated with them had competitive advantages in the green tea clusters in Japan. These included leading the quality frontier, the ease of access to public goods to encourage producers' incentives, and enhanced demand conditions.

Finally, there are tasks remaining for the next steps. One task is an analysis of the effect on green tea clusters by introducing the existence of an increase in bottled tea consumption. The drinking style of green tea has changed from the traditionally one of brewing in a pot to the bottled one. According to the Survey of Household Economy, the expenditures on brewed green tea per household have declined from 6,810 to 4,224 yen. On the other hand, spending on bottled tea rose from 3,668 to 5,889 yen. Teramoto (2001) pointed out that consumers prefer a healthy beverage and, at the same time, favor the convenience of bottled tea rather than the trouble of brewing tea. Akune (2011) indicated this impact on domestic green tea industries by using an input–output analysis. However, it is necessary to analyze the impact on regional green tea clusters, and especially to focus on the competitive advantages that differ compared to the advantages described in this paper. Another task is a quantitative analysis of the competitive advantages among green tea clusters. In this paper, several factors were found as differences among the clusters of several prefectures. It is necessary to indicate concretely the competitive advantages, and the determinants for making a plan for the clusters that brings high productivity and competitiveness.

Acknowledgments This work was supported by JSPS KAKENHI Grant Number 25850154.

References

Akune Y (2009) The empirical analysis on agglomeration and location choice of food industries [Syokuhin-sangyo no Sangyo-syuseki to Ritti-sentaku ni ka-n-su-ru Jissyou-bunseki]. Tsukuba Shobo, Tokyo [in Japanese]

Akune Y (2011) Green tea clusters by green tea manufacturers and the economic effect of consumption change [Seitya-gyo wo tyu-sin-to-si-ta "Ryokutya Kurasuta" to Jyuyou-henka ni to-mo-na-u Ryokutya-kanren-sangyou he-no Eikyou-do-shisan]. In: Kiminami L, Nakamura T (eds) Food security and industrial clusters in Northeast Asia [Hokuto-asia no Syokuryo-anzen-hosyou to Sangyou kurasta]. Agriculture and Forestry Statistics Publishing, Tokyo, pp 146–161 [in Japanese]

Akune Y, Tokunaga S (2003) Agglomeration in food industries and within food firms in Japan [Kokunai-ni-okeru Syokuhin-sangyou no sangyousyuuseki no henka to kigyounai deno sangyousyuuseki]. J Rural Econ [Nihon Nogyo Keizai Gakkai Ronbunshu] 2003:326–328, Special Issue [in Japanese]

Ellison G, Glaeser EL (1997) Geographic concentration in U.S. manufacturing industries: a dartboard approach. J Polit Econ 105(5):898–927

Kageyama M, Tokunaga S, Akune Y (2006) Agglomeration of wine industry and a wine cluster: the case of Katsunuma region in Yamanashi [Wain-Sangyo no Syuseki to Wain-Cluster no Keisei]. J Food Syst Res 12(3):39–50 [in Japanese]

Morishima T (2012) Network analysis of food industry clusters [Syokuryo-sangyo Kurasuta no nettwaku Kouzou-bunseki]. Nourin Toukei Kyoukai, Tokyo [in Japanese]

Porter ME (1998) On competition. Harvard Business School Press, Boston

Saito O (2007) Food industry clusters and regional brands [Syokuryou-Sangyou Kurasuta to Tiiki-Brand]. Rural Culture Association [Nou-san-gyo-son Bunka Kyoukai], Tokyo [in Japanese]

Teramoto Y (2001) Quantitative analysis of the recent trends of green tea consumption. Keizaigakuronkyu, Kwansei Gakuin Univ 55(3):41–72 [in Japanese]

Chapter 9
Food and Health-Related Industry Clustering in Niigata Prefecture

Empirical Analysis on the Cognitive Aspects of Corporations

Lily Kiminami and Shinichi Furuzawa

Abstract This study aims to clarify empirically the policy issues for regional innovation focusing on the "health-related industry cluster" in Niigata Prefecture, Japan. More specifically, it clarifies the situation of the "health-related industry cluster" and the impacts of spatiality of knowledge flow on the regional innovation system based on an interview survey of the intermediate organizations and a questionnaire investigation targeting the members of the cluster. Additionally, it identifies the relationship between inter-firm cognitive distance and innovation creation based on the analysis of annual security reports of three leading companies in the cluster using text mining method. Finally, it draws policy implications for the strategy of regional innovation based on the analytical results and points out the tasks for future research.

Keywords Regional innovation • Food and health-related industry clustering • Cognitive distance • Knowledge flow

1 Introduction

In recent years, innovation theory has opened up a new horizon by absorbing traditional learning theory and system theory based on cognitive science. This new innovation theory pointed out the importance of a social scene focusing on the structure of individual and collective consciousness. Furthermore, recent studies related to knowledge and innovation implied that organizational linkages which dominate the multilayered spatial dimension (from local to global) determined the possibilities of consecutive induction of innovation by circulating the knowledge spillover out of

---

L. Kiminami (✉) • S. Furuzawa
Institute of Science and Technology, Niigata University, Niigata, Niigata, Japan
e-mail: kiminami@agr.niigata-u.ac.jp; furuzawa@agr.niigata-u.ac.jp

the industrial agglomeration, in addition to skilled labor and the strength of institutions to support industry, such as in culture, society and politics.

Meanwhile, in Japan, society has been undergoing a rapid aging, birthrates have been falling, and the issues for solving the problems of health and improving the quality of life have become diverse and complex. For realizing the sustainability of regional development, the issue of innovation creation has become urgent, with the progress in globalization and the tightening constraints on the environment and resources (APO 2011).

Therefore, the purpose of this study is to clarify empirically the policy issues for regional and local innovation focusing on the "health-related industry cluster" in Niigata Prefecture, Japan, which is thought to have entered a growth phase. More specifically, it will clarify the situation of the "health-related industry cluster" based on an interview survey of the intermediate organizations of the cluster, and clarify the impacts of the spatiality of knowledge flow on the regional innovation system based on a questionnaire investigation targeting the members of the cluster. Additionally, it will identify the relationship between the inter-firm cognitive distance and innovation creation based on an analysis of the annual security reports of three leading companies in the cluster using text mining method. Finally, it will draw up policy implications for the strategy of regional innovation based on the analytical results.

2 Literature Review on the Spatiality of Knowledge Flow and Cognitive Aspects of Corporations

Knowledge and innovation are created at multiple levels (from local to global) throughout the network in variously spatial dimensions. While documented explicit knowledge is in general capable of long-distance transmission, the transmission of tacit knowledge embodied in people necessitates face-to-face contact. Therefore, in light of the different tacit and explicit features within knowledge, Nonaka and Takeuhi (1995) discussed the importance of "*ba*" (social scene) for knowledge creation. In addition, using the classification of explicit knowledge and tacit knowledge, Asheim and Coenen (2005, 2006), and Asheim and Gertler (2005) extended the discussion to the existence of the knowledge bases of "synthetic" and "analytical".

On the other hand, as for studies on the relationship between innovation and the cognitive aspects of corporations,[1] analyses have been conducted focusing mainly on: (i) patent information and (ii) knowledge flow from an economic geography approach (Matsubara 2013; Mizuno 2011). In addition, some studies analyzing the cognitive innovation of corporations from the viewpoint of organizational learning theory (Kida 2007) indicated that the actual condition of organizational learning

---

[1] Regarding studies of the cognitive aspects of corporations, some studies in the field of economic sociology have been conducted. See also Stark (2009) for the concept of dissonance and the importance of corporations which have the capacity for reflective cognition.

might be effectively grasped from the "change of organizational knowledge structure" by applying the method of text-mining, in which a certain relationship between the cognitive change and the management performance of a company was clearly shown.

Furthermore, the global-knowledge society creates its sources of growth from diversity at multiple levels, where a structure with a higher level of knowledge coordinates global competition and cooperation. The difference in each particular culture and system is cross-referenced as "intellectual resources" for new knowledge creation, and cognitive diversity serves as a new (and moreover, inexhaustible) resource for a sustainable society (Tayanagi 2007). In spite of the importance of the cognitive aspects of regional innovation and cluster formation, empirical analysis thereof is fairly scarce. Therefore, clarifying the impacts of cognitive diversity, which is measured as the cognitive distance for the cognitive innovation of a corporation, is critically important for effective strategy for regional innovation.

3 Case Study from Industry Clustering in Niigata

3.1 Policy for the Health-Related Industry Cluster

In Niigata Prefecture, in order to make the business of health-related industry high value-added alongside the aging of society, "A new vision for health, welfare, and the medical industry (February 2006 release)" is promoted by the government, and named "Health-Related Industrial Business" covering broad industries, such as agriculture, forestry and fisheries, the food industry, the machinery industry, service industries, the tourist industry, and the healthcare industry, in connection with health, welfare, and medical treatment.

The relationship between the main organizations and concrete projects is composed of two parts: (i) location policy with crossover administrative organization; and (ii) policy for product development and a new business model by cross-industrial cooperation. Emphasis is put on the latter especially. Moreover, the main objects of the support from cross-industrial cooperation are small and medium-sized enterprises (SMEs) located in the prefecture, through information exchange with supporting organizations, public research organizations, and universities, etc.

The Policy for Health-Related Industry has shifted from the planting/training stage (the first stage, fiscal years 2006–2008) to the marketing/expansion stage (the second stage, fiscal years 2009–2011). In March 2011, 15 cases were selected as a recommended business model from a viewpoint of market appeal power (best practice). However, the present situation is that there are few corporations which follow the best practices. According to an interview with the person in charge of the Industry Promotion Division of Niigata Prefecture, two points are stressed as the causes. First, there is a lack of entrepreneurship in the region in the business environment. Second, there are insufficient funds for the private sector, such as the funding for venture businesses, especially for angel investors.

Table 9.1 Actors and channels of knowledge flow

	Actors		
	Member corporations of JHBF	Non-member corporations	Administrations
Document	Annual security reports, etc.	Annual security reports, etc.	Selection of best practices
Oral	**Sectional meeting activity**	Uonuma meeting	Uonuma meeting
	Uonuma meeting		
Electronic	E-mail		
Products	Sectional meeting activity (Food tasting, etc.)		
Cooperation			
Embodied to person	**Academy for human resource development**	**Academy for human resource development**	

Source: Kiminami and Furuzawa (2015)
Notes: "Electronic" is information-sharing by electronic media. "Cooperation" is a joint research project, etc.

Corresponding to such a situation, the support policy of Niigata Prefecture for innovation has shifted to personnel training since 2012. The representative case is to give a commission to the Japan Health Business Federation (JHBF) for running an "Academy for Human Resource Development"[2] from fiscal year 2012.

Table 9.1 summarizes the actors and channels of knowledge flow centering on JHBF. The knowledge exchange between the member companies of the JHBF has been realized directly through the activity of sectional meetings, while the knowledge exchange with non-member companies has been expected indirectly through the "Uonuma Meeting" and the "Academy for Human Resource Development", etc. However, the knowledge flow between member companies still has a one-way characteristic, and the incentive mechanism for creating new knowledge is ineffective, although it functions to some extent for the combination of existing knowledge. Therefore, it is important to build a system for promoting the interaction of the knowledge flow among entities, including the non-member companies and administrations, instead of holding sporadic meetings.

3.2 Quantitative Analysis

To clarify the impact of cluster formation on the competitiveness of companies, regression analysis was undertaken using the following measurement equation.

$$Yj = \alpha + \Sigma \beta i\ Xij + \gamma_1 Cj + \gamma_2 Sj + \gamma_3 Gj + \varepsilon j \tag{9.1}$$

---

[2] Mainly based on the author's interview investigation of the JHBF on 8 November 2012.

Here, *Y* is the performance of companies (profit ratio of sales: 2006–2008, year average), *Xij* is the profile of companies (Years of Operation, Number of Employees, Scale of Capital Amount), *Cj* is a Cluster Dummy (Business Partner with a Member: 1, None: 0), *Sj* is a Sector Dummy (Manufacturing: 1, Non-manufacturing: 0), *Gj* is a Region Dummy (Niigata Prefecture: 1, Others: 0). The Cluster Dummy (*Cj*) is used to test the effect of cluster formation on company performance. If the sign of the parameter *Cj* is positive, it can be judged that this cluster has an upgrading effect on the competitiveness of companies. Moreover, if the sign of *Gj* is positive and significant, this cluster has the effect of geographic accumulation. In other words, from the viewpoint of industrial location, it can be considered that forming a healthcare business cluster in Niigata holds economic rationality. Table 9.2 shows the result of the measurement by the least squares method (enterprises that don't obtain information on attributes and performance are excluded), in which the following four points are clarified:

1. Each of the variables of corporate profile is not significant;
2. The Cluster Dummy is not significant;
3. The sign of the Region Dummy is positive and statistically significant;
4. The sign of the Sector Dummy is negative and statistically significant.

Therefore, it can be explained that the relationship between companies' profiles and performance is not linear, and the economic effect of this cluster is almost 0 at the first stage. Moreover, the effect of geographic agglomeration of enterprises is confirmed from the positive sign of the Region Dummy, and it can be judged that there is economic rationality for the location. In addition, the negative sign of the Sector Dummy indicates that the manufacturing sector, especially the food sector, has reached the mature stage of the product life cycle.

3.2.1 Basic Structure of the Network

To grasp the overall structure of this network, the basic indicators of this network are summarized in Table 9.3. Moreover, when the degree of distribution of the food sub-cluster in the network is compared with a random network, it is understood that the network of the food sub-cluster has a shape close to the random network. It was not possible to confirm the network of the small-world type. However, nodes which have large links exist.

3.2.2 Extraction of the Communities

Next, to group the companies on the network, we use the "Modularity *Q*" index proposed by Newman (2004) in the same way as Sakata et al. (2007). Modularity *Q* can be shown by the following expression:

$$Q = \sum_{s=1}^{N_m} \left[(l_s / l) - (d_s / 2l)^2 \right] \qquad (9.2)$$

Table 9.2 Regression analysis on performance of companies (least squares method)

Explained variable			Profit ratio of sales (2006–2008 average)		
Explanatory variables			Coefficient	t-value	P-value
Constant			−0.055	−1.29	0.205
Profile		ln(Years of operation)	0.004	0.45	0.655
		ln(Capital amount)	0.001	0.23	0.821
		ln(Number of employees)	0.001	0.16	0.870
Dummy variable		Cluster dummy (business partner with a member: 1, none: 0)	0.001	0.09	0.929
		Sector (manufacturing: 1, non-manufacturing: 0)	−0.019	−2.23	0.032**
		Region (Niigata prefecture: 1, other prefectures: 0)	0.047	2.97	0.005***
Adjusted R^2			0.222		
Number of observations			43		
F-value			3.003		
Significant F-value			0.018**		
Basic statistical values					
		Average	S.D.	Min.	Max.
ln(Years of operation: Year)		3.813	0.563	1.792	4.883
ln(Capital amount: Thousand Yen)		13.197	3.232	9.210	20.332
ln(Number of employees)		5.540	2.313	1.386	10.504
Cluster dummy (business partner with a member: 1, none: 0)		0.442	0.502	0	1
Sector (manufacturing: 1, non-manufacturing: 0)		0.837	0.374	0	1
Region (Niigata prefecture: 1, other prefectures: 0)		0.349	0.482	0	1
Profit ratio of sales (2006–2008 average): profit/ratio		0.009	0.028	−0.074	0.091

Note: "***", "**", "*", indicate statistically significance at the 1 %, 5 %, and 10 % levels, respectively

Table 9.3 Basic indicators of network structure

Indicator			Value
Distance	Diameter	T: max d_{ij}	5
	Average distance	$2 \cdot \sum d_{ij}/N \cdot (N-1)$	2.743
Cohesion	Density	$2 \cdot L/N \cdot (N-1)$	0.058
	Transitivity		0.073
Number of links		L	254
Number of nodes		N	94

Source: Kiminami et al. (2012)

Here, N_m is the number of modules, l_s is the number of links in module s, l is the number of links in the network, and d_s is the number of links held by the nodes in module s. The right clause 1 is an actual measurement value of the probability that links exist between nodes in the module. Clause 2 is a theoretical value of the ratio of the links in the module when assuming a random network. The computational method is as follows. First of all, the connection of nodes is advanced so that the Q value may become the maximum. The number of modules that become Q_{max} immediately before the ΔQ value takes a negative value is calculated, and the module that each node (enterprise) belongs to is extracted. Figure 9.1 shows the numerical results concerning the change of the Q value.

Based on the above-mentioned results, a graph of the network is drawn, and the shape and the color of the nodes are differentiated according to the type of business sector and the community (Fig. 9.2).

3.2.3 Extraction of Core-Companies

Next, two indices concerning Centrality (Degree Centrality, and Betweenness Centrality) were calculated to analyze the feature of hub companies in the network, and the results are shown in Table 9.4.

With regards to the Degree Centrality, Bourbon, Kameda Seika and Iwatsuka Seika, which are members of the cluster, are the top three companies. These three companies are located in Niigata Prefecture and are the leading companies in this area. Also, these three companies are advancing into foreign countries through direct investment in those countries. Except for manufacturing industry, Mitsubishi, Kamiyama Bussan, and Mitsui, which are classified into the wholesale and retail trade, are top in the Food Sector. Moreover, Kamiyama Bussan is an enterprise located in Niigata Prefecture, and has a role of a circulation hub in the local food network. As for the Betweenness Centrality, it has an almost similar tendency.

The common characteristics of these companies are being global–local companies that have a strategy of expanding into overseas markets on the one hand, and selling their products to domestic metropolitan markets by taking advantage of their location locally, on the other. Moreover, Kameda Seika and Iwatsuka Seika are placed in a similar position in the network, and the degree of competition is stiff.

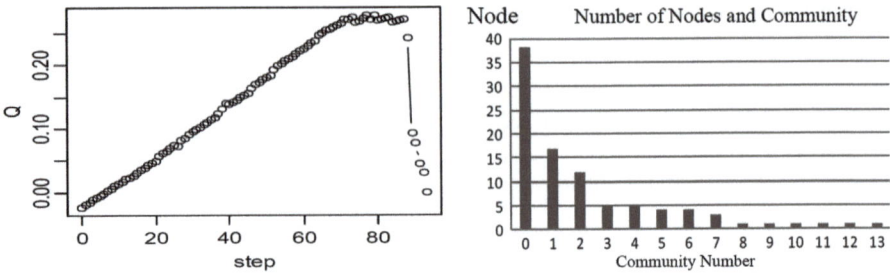

Fig. 9.1 Results of modulation generation (Source: Kiminami et al. 2012)

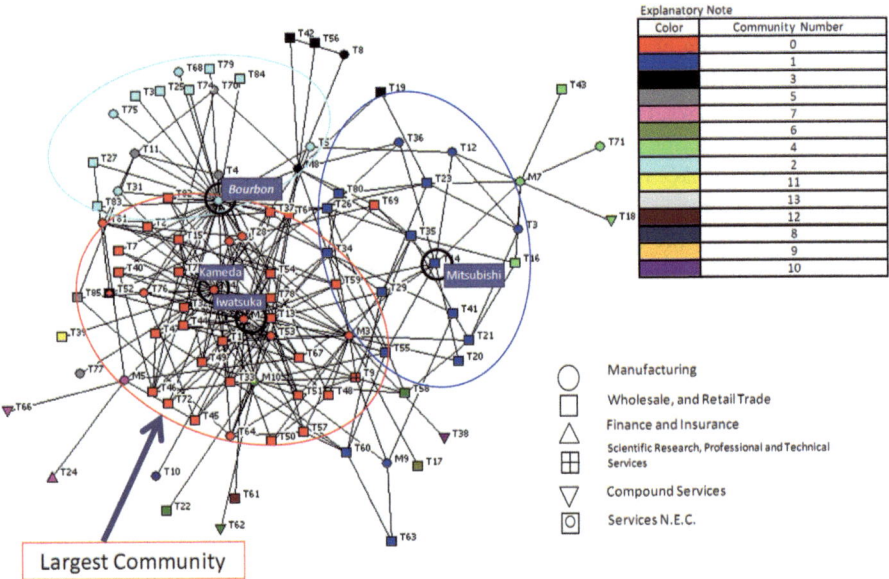

Fig. 9.2 Network space of food sub-cluster (Source: Kiminami et al. 2012)

Additionally, these companies are the leading enterprises in the food manufacturing industry, and their corporate performance is generally excellent. However, the competitive advantage of a company is decided both by its position in the industry and the attractiveness of the industry. Therefore, participating in the formation of a health-related industry cluster in Niigata is an important business strategy for the companies not only for strengthening their own competitiveness but also for improving the attractiveness of the industry in order to develop the food sector into a new industry.

9 Food and Health-Related Industry Clustering in Niigata Prefecture

Table 9.4 Centrality index and core companies

Name of companies	Industrial sector	Region	Member of convention M: Member; T: Business partner	Centrality		Module
				Degree centrality	Betweenness centrality	Community number
Bourbon	Manufacturing	Niigata	M	34	1,139.245	2
Kameda seika	Manufacturing	Niigata	M	32	905.830	0
Iwatsuka seika	Manufacturing	Niigata	M	28	529.153	0
Echigo seika	Manufacturing	Niigata	M	19	466.510	0
Mitsubishi	Wholesale & retail	Tokyo	T	18	228.008	0
Maruei seihun	Manufacturing	Niigata	M	13	503.786	3
Yamazaki jyouzou	Manufacturing	Niigata	M	12	332.368	6
Kamiyama bussan	Wholesale & retail	Niigata	T	11	153.272	0
Uchiyama-tousaburo-shouten	Manufacturing	Niigata	T	10	90.838	0
Mitsui	Wholesale & retail	Tokyo	T	9	244.270	1
Ryoshoku	Wholesale & retail	Tokyo	T	9	167.947	2
Niigata kenbei	Wholesale & retail	Niigata	T	9	98.900	0

Source: Kiminami et al. (2012)

3.3 Questionnaire Survey Targeting the Members of the Cluster

Next, we conducted a questionnaire investigation of JHBF members in order to clarify the role of the federation and the actual situation of innovation creation through the activity of the cluster. The questionnaires were sent by e-mail through the staff of the federation on 23 August 2013, and collected by e-mail and fax up until 11 September 2013. The distributed number was 116 and the number of respondents was 12 (collection ratio: 10.3 %). The main items of the investigation were the profile of the respondents, the required role of the federation, the innovation activity in the cluster focusing on the types of innovation (product, process, organization, and marketing) and collaboration, and objectives, etc.

First, as for the role of the federation, "g. Creating a new business model by communicating with different industries" was selected most (8 enterprises among 12) followed by "a. Searching for partners for co-developing new products, services and technologies", "b. Searching for a sales destination for one's company's products and services", "f. Acquiring information on related industries", and "h. Communicating with outstanding entrepreneurs" (7 enterprises). It may be said that the member enterprises consider the federation as a platform not only for searching for business partners, but also for seeking possibilities of new business by communicating with different industries and entrepreneurs with the emphasis on accessing information rather than transmitting information. However, only one enterprise selected the response for human resources, despite the fact that the federation has been focusing on the cultivation of human resources in recent years.

Second, as for the realization of innovation in health-related business over the previous 3 years, each respondent had created one or all the types of innovation in terms of product, process, organization, and marketing. Additionally, innovation through organizational collaboration was conducted as well. Especially interesting is that those that attained organizational innovation are the enterprises that have both conducted collaboration with universities and attained marketing innovation.

Third, as for the objectives of innovation activities, the large numbers for "Upgrading and expanding one's company's products and services", "Improvement of the quality of products and services" and "Entering and expanding of the market" mean that enterprises which achieved innovation creation have conducted organizational collaboration and reached out to the market based on a strategy of products and services. As for the experience of cooperation with enterprises within the prefecture, however, "Providing information" is the highest (five enterprises), followed by "Technological offerings" and "Proposing business models" (three enterprises). On the other hand, the number of enterprises without experience of cooperation with other entities is still large (six enterprises).

The following actual situation is highlighted from the results of the investigation. The activities of the cluster mainly focus on knowledge-sharing through introducing new products and practices among small and medium-sized enterprises; the actual achievement of a new business model and market creation is scarce, and; there is a large number of sleeping actors in the federation.

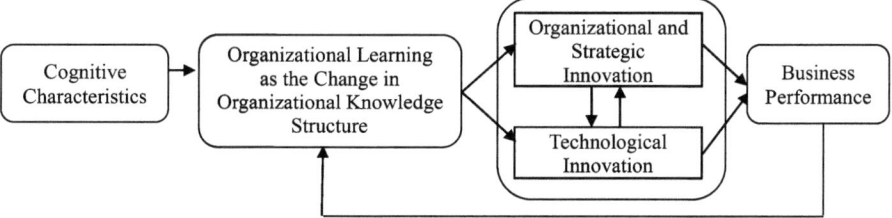

Fig. 9.3 Basic model of cognitive innovation of corporation (Source: Kiminami and Furuzawa 2015). Note: Revision based on Kida (2007, p. 32) Tables 1–10.

3.4 Analysis of the Cognitive Innovation of Corporations

3.4.1 Data and Basic Model Used in This Research

Next, we analyze how the cognitive innovation of corporations (the organizational, technical and strategic innovation accompanying organizational learning as a change in organizational knowledge structure) occurs under the formation of a Health-Related Industry Cluster conforming to the following basic model (Fig. 9.3). The actual condition of cognitive innovation of three core corporations in the food sub-cluster of the Health-Related Industry Cluster are analyzed by using the method of text-mining. The data used for the analysis are from documents (the 9 years from March 2004 to March 2012, fiscal years) published in the "Annual Security Report" and the target text "the issue which should be coped with" is considered as the cognitive information about management issues in this research, linking with "organizational learning as a change in organizational knowledge structure" and "organizational and strategic innovation", "continuous and planned searching processes based on the target of the corporation for optimally solving problems" (Ansoff 1965). The management concept and strategy of each corporation are read from the collocation maps. Furthermore, how the change in organizational knowledge structure affects organizational innovation and technological innovation, and what the impact is of the organizational and strategic innovation and technological innovation on the business performance of the three corporations, are clarified, referring to the situation for patent acquisition and the recurring profit margin.

3.4.2 Important Words and Collocations

In this research, "Important Concepts" are extracted by using the "term frequency-inverted document frequency (tf-idf)". Among the top 15 important concepts, we selected the innovation-related concepts as "Remarkable Concepts". First, the remarkable concepts are extracted from the important concepts, where: those of corporation B are "development", "health", and "research"; those of corporation K are "brand" and "profits", and; that of corporation I is "profits".

Furthermore, in order to grasp the relationship between the cognitive change and innovation of corporations, the collocations are analyzed by making "development", "brand", and "profits" the main words. The main word of corporation B was set as "development". The top collocation is "goods" throughout this period. Although "research" had appeared as a collocation from 2004 to 2006, "technology" appeared from 2008 to 2011. It has been clarified that the cognition of corporation B concerning "development" is based on "product development", and the importance of development has shifted from "research" to "technology".

As for corporation K, the main word was set as "brand". In 2005 and 2006, "trust" was the top collocation and "value" was ranked top from 2007 to 2012. Moreover, "stockholder" and "company" appeared continuously from 2008 to 2012. It turns out that the cognition of corporation K concerning "brand" has shifted from "trust" to "value" while changing the direction for the construction of a corporate brand by being more conscious about stockholders.

As for corporation I, the main word was set as "profits". In 2007 and afterwards, "stockholder" and "cooperation" appeared. Therefore, it is thought that the cognition of corporation I concerning "profits" is continually aware of the "stockholder".

3.4.3 The Actual Situation of Technological Innovation: Technological Fields and Forms of Patents

Table 9.5 summarizes the technological fields and application form for the patents of the three core corporations. Regarding application-based patents, corporations B and K not only have them in the field of "food", but also in "medical goods", and with organizational cooperation at a constant rate. Regarding registration-based patents, corporation K has the most in terms of number (30 cases) and is the only one in the technological field of "medical goods" via industry–university cooperation.

Therefore, according to the knowledge-based approach, the three corporations belong to the integrative knowledge-based industry requiring the application and combination of existing knowledge. However, corporations B and K are shifting to the analytic knowledge-based industry requiring advanced knowledge and technology. As for the research-and-development system, corporation B conducts product development paying attention to the nutritional features of food in cooperation with domestic universities. Additionally, corporation K is conducting product development in cooperation with local universities by focusing on the health functions of rice. Although the rate of patent registration (registration/application number) of corporation I is 100, there is little joint patent application in cooperation with a university or a research institution which would have highly original technology, and it has withdrawn its business from the overseas market.

Table 9.5 Technological field and form of patents (application basis). Unit: Cases (%)

Classification			Application Basis			Registration Basis		
			B	K	I	B	K	I
Technological field (Top IPC)	1	Agriculture and fishery	1 (2.6)	2 (3.0)	0 (0.0)	0 (0.0)	0 (0.0)	0 (0.0)
	2	Food	20 (51.3)	49 (74.2)	6 (100.0)	9 (75.0)	23 (76.7)	6 (100.0)
	3	Personal and household articles	0 (0.0)	2 (3.0)	0 (0.0)	0 (0.0)	1 (3.3)	0 (0.0)
	5	Medical goods	6 (15.4)	5 (7.6)	0 (0.0)	0 (0.0)	1 (3.3)	0 (0.0)
	11	Packing, container, storage, heavy machine	11 (28.2)	2 (3.0)	0 (0.0)	3 (25.0)	2 (6.7)	0 (0.0)
	12	Inorganic chemistry, fertilizer	0 (0.0)	1 (1.5)	0 (0.0)	0 (0.0)	0 (0.0)	0 (0.0)
	13	Organic chemistry, pesticide	1 (2.6)	0 (0.0)	0 (0.0)	0 (0.0)	0 (0.0)	0 (0.0)
	14	Macro molecule	0 (0.0)	1 (1.5)	0 (0.0)	0 (0.0)	0 (0.0)	0 (0.0)
	16	Biotechnology, beer, liquor, carbohydrate industry	0 (0.0)	2 (3.0)	0 (0.0)	0 (0.0)	2 (6.7)	0 (0.0)
	27	Measurement, optics, and photocopy machine	0 (0.0)	2 (3.0)	0 (0.0)	0 (0.0)	1 (3.3)	0 (0.0)
Form of applicants		Singularity	29 (74.4)	53 (80.3)	6 (100.0)	10 (83.3)	24 (80.0)	6 (100.0)
		Industry-academia	2 (5.1)	5 (7.6)	0 (0.0)	0 (0.0)	1 (3.3)	0 (0.0)
		Industry-industry	5 (12.8)	8 (12.1)	0 (0.0)	2 (16.7)	5 (16.7)	0 (0.0)
		Industry-academia & industry-industry	3 (7.7)	0 (0.0)	0 (0.0)	0 (0.0)	0 (0.0)	0 (0.0)
Total			39 (100.0)	66 (100.0)	6 (100.0)	12 (100.0)	30 (100.0)	6 (100.0)

Source: Kiminami and Furuzawa (2015)
Notes: Authors' Calculation using the data from IPDL (http://www2.ipdl.inpit.go.jp/begin/be_logoff.cgi?sTime=1378347074) (Retrieved: 5 September 2013). The classification of the Technological Field is based on Goto and Motohashi (2007, p. 1433)

3.4.4 Trends in Business Performance: Number of Patents, Recurring Profit Margin (RPM) and Collocations of Words

Figure 9.4 shows the relationships between the number of patents, RPM and cognitive importance of "development", "research" and "human resources", which are related concepts of business performance.

For corporation B, the cognitive importance of the concepts has a negative correlation with the number of patent applications and a positive correlation with the number of patent registrations (a time lag between the number of patent applications and registrations). The number of patent registrations had increased along with an increase in the cognitive importance of the concepts of "research" and "human resources" before the business performance upturn in 2008 from a period of low performance from 2000. Therefore, it is inferable that the upturn of business performance is led by organizational cognitive innovation.

As for corporation K, even though there isn't any correlation between the cognitive importance of the concepts and the number of patents, the cognitive importance of "development" and "research" has increased after 2007 (the year with the lowest number of patents), for which it is supposed that changes have arisen in the organizational knowledge structure, and it is also supposed as being successful in innovation because it has come to have positive earnings since 2000.

As for corporation I, the cognitive importance of "human resources" and "development" sharply decreased in 2005 and 2006, respectively, while there was no correlation between the cognitive importance of "development", "research" and "human resources" and the number of patents. Moreover, changes in the organizational knowledge structure which leads to cognitive innovation might have barely been induced for corporation I since its RPM was kept at a constant level by a repetition of increases and decreases.

Based on the above results, the relational model between the organizational knowledge structure and innovation of each corporation can be drawn as follows (Fig. 9.5). Corporations B and K transformed their knowledge structure through organizational learning corresponding to the change in cognitive characteristics and formed their new strategies respectively. As a result, business performance is improved by organizational and technological innovation. However, the capability of corporation K for inducing the changes in knowledge structure through organizational learning, which is considered an interpretation of the change in its cognitive characteristics, is a little weak. On the other hand, although corporation I changed its strategy, quickly corresponding to the change in cognitive characteristics inducing a short-term and evolutionary technological innovation, there was no cognitive innovation through organizational learning. Therefore, for corporation I a model transformation toward organizational innovation through organizational learning for achieving a constantly high business performance is called for.

9 Food and Health-Related Industry Clustering in Niigata Prefecture

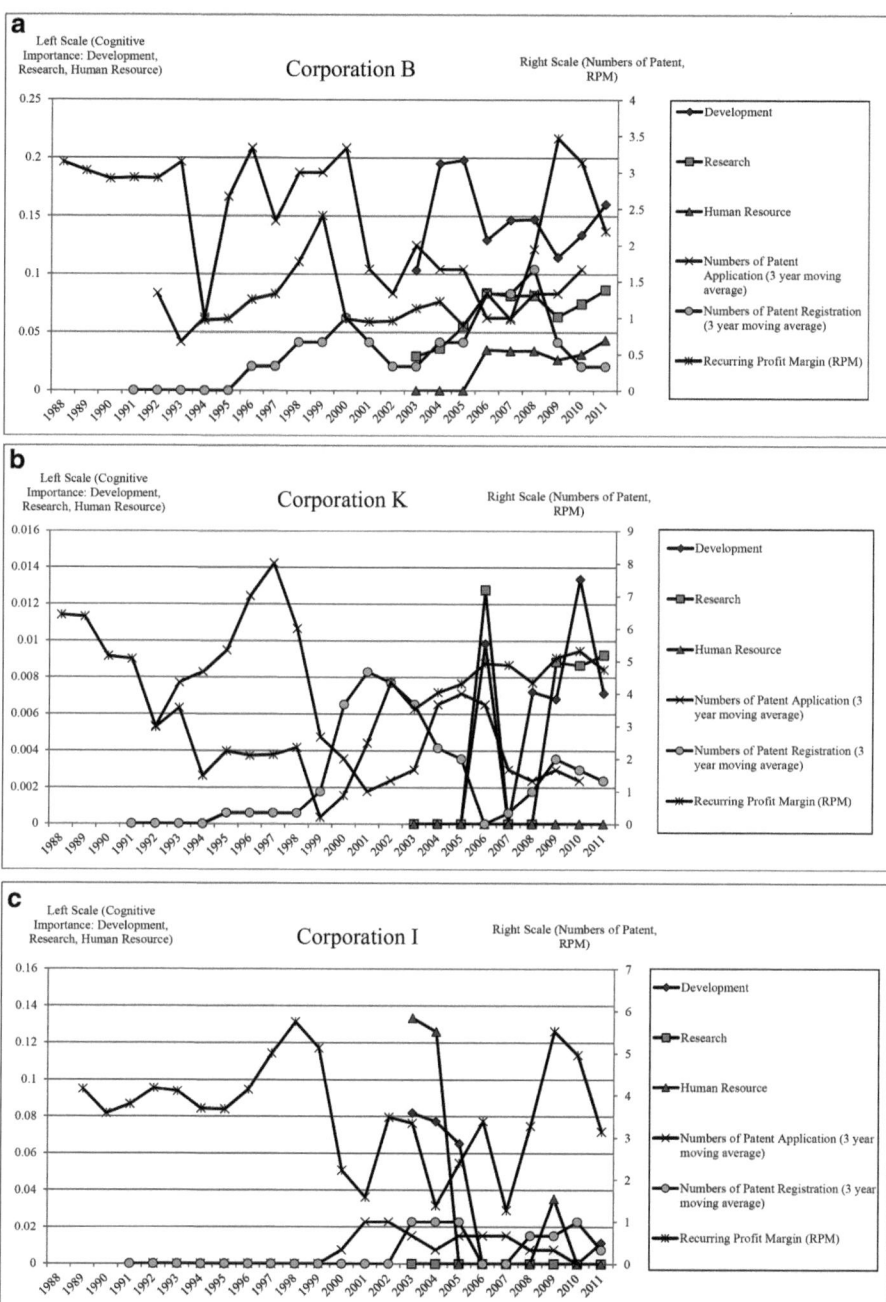

Fig. 9.4 (**a**) Relationship between cognitive importance and innovation (Corporation B) (Source: Kiminami and Furuzawa 2015). (**b**) Relationship between cognitive importance and innovation (Corporation K) (Source: Kiminami and Furuzawa 2015). (**c**) Relationship between cognitive importance and innovation (Corporation I) (Source: Kiminami and Furuzawa 2015)

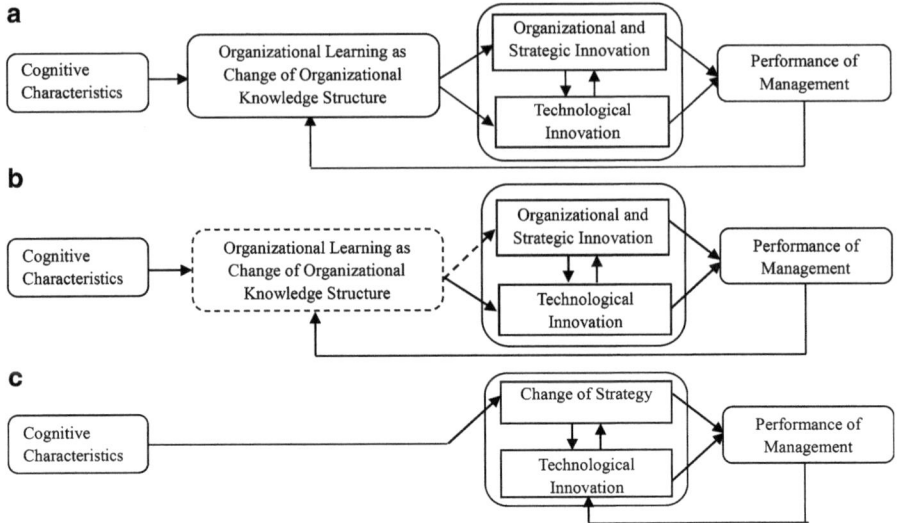

Fig. 9.5 Model of cognitive innovation of the three corporations. (**a**) Corporation B, (**b**) Corporation K, (**c**) Corporation I. (Source: Kiminami and Furuzawa 2015)

4 Conclusions

In conclusion to the aforementioned analytical results, the cognitive change of subjects as organizational learning may induce innovation has been confirmed empirically. First of all, an effectively formed cluster would influence knowledge creation among the subjects which are thought to be the cause of innovation. Second, the results of our questionnaire survey showed that the member enterprises of the cluster consider JHBF as a platform not only for searching business partners but also for seeking the possibilities of new business by communicating with different industries and entrepreneurs. Each respondent has created one type or all types of innovation for product, process, organization and marketing. Third, the results of our analysis on the cognitive innovation of leading companies in the cluster clarified that the cognitive innovation of corporations (the organizational, technical and strategic innovation accompanying organizational learning as the change in organizational knowledge structure) occurred with the formation of a health-related industry cluster, such as corporations B and K, which have succeeded in innovation in the domain of new technology and markets. Fourth, the main actors of the federation, such as food-manufacturing industries, started to deepen the networks of enterprises both vertically and horizontally through the momentum of related promotional policies and the activities of the federation. However, the linkages with the local institutions in education and research are a little weak in spite of the federation being required to revitalize the sub-system of knowledge creation and dissemination through the "Academy for Human Resource Development (HRD)".

As tasks for future research, the following points need to be mentioned. First of all, the relations between cognitive features and changes and innovation creation can be further clarified through interviews at the leading companies. Moreover, the effects of the cluster can be evaluated through comparing the cognitive changes of the entities through organizational learning between the interior and exterior of the cluster. Finally, careful case studies of both successful and unsuccessful innovation creation should be put into our agenda.

References

Ansoff HI (1965) Corporate strategy. McGraw-Hill, New York
APO (Asian Productivity Organization) (2011) Population aging and productivity in Asian countries. APO, Tokyo
Asheim BT, Coenen L (2005) Knowledge bases and regional innovation systems: comparing Nordic clusters. Res Policy 34(8):1173–1190
Asheim BT, Coenen L (2006) Contextualising regional innovation systems in a globalising learning economy: on knowledge bases and institutional frameworks. J Technol Transfer 31(1):163–173
Asheim BT, Gertler M (2005) The geography of innovation: regional innovation systems. In: Fagerberg J, Mowery D, Nelson R (eds) The Oxford handbook of innovation. Oxford University Press, Oxford, pp 291–317
Goto A, Motohashi K (2007) Construction of a Japanese patent database and a first look at Japanese patenting activities. Res Policy 36(9):1431–1442
Kida M (2007) Cognitive study on organizational innovation: visualization of cognitive change and knowledge and introduction of text mining to organizational science. Hakuto-Shobo Publishing, Tokyo [in Japanese]
Kiminami L, Furuzawa S (2015) Theoretical and empirical study on regional and local innovation: focusing on the health-related industry cluster in Niigata, Japan. Stud Region Sci 44(4):495–515
Kiminami L, Furuzawa S, Kiminami A (2012) Analysis of network structure of a food sub-cluster: case study of the health-related industry clustering in Niigata prefecture, Japan. Stud Region Sci 41(4):1055–1074 [in Japanese]
Matsubara H (2013) Industrial clusters and theory of regional innovation. In: Matsubara H (ed) Cluster policy and regional innovation in Japan. University of Tokyo Press, Tokyo, pp 3–25 [in Japanese]
Mizuno M (2011) Economic space of innovation. Kyoto University Press, Kyoto [in Japanese]
Newman MEJ (2004) Fast algorithm for detecting community structure in networks. Phys Rev E 69(6):066133
Nonaka I, Takeuhi H (1995) The knowledge-creating company: how Japanese companies create the dynamics of innovations. Oxford University Press, Oxford
Sakata I, Kajikawa Y, Takeda Y, Hashimoto M, Shibata N, Matsushima K (2007) Network dynamics in the twelve regional clusters, RIETI discussion paper series 07-J-023
Stark D (2009) Heterarchy: the organization of dissonance. The sense of dissonance: accounts of worth in economic life. Princeton University Press, Princeton, pp 1–34
Tayanagi E (2007) A trend of 'cognitive approach' in social sciences: in association with policy implementation. Cognitive Stud, Bull Jpn Cognitive Sci Soc 14(1):60–73 [in Japanese]

Chapter 10
Agricultural Industry Clusters in China

Lily Kiminami and Akira Kiminami

Abstract The purpose of this study is to assess the potential of clustering in the development of agriculture and rural communities in China. First, case studies are used to clarify issues involving cluster initiatives and business environments. The second method consists of quantitatively analyzing the relationship between business collaborations among companies and economic performance. The results of the analyses of agricultural clusters in China in this study suggest that clusters do achieve certain positive results in the way of development in the agricultural sector and rural communities. However, it also suggests that a shift from intra-industry to inter-industry business collaboration will be necessary for the sustained development of those clusters.

Keywords Agricultural industry cluster • Cluster initiative • Business collaboration • Development of agriculture and rural community

1 Introduction

Even a cursory examination of the agricultural sector and rural communities in China today reveals evidence that industrial clustering is taking root in that country (Wang 2011). Structural changes are clearly taking place in the agriculture and food industries, led by the so-called "dragon-head" (leading) enterprises that serve as integrators of production, sales, distribution and processing functions in the sector. The major dragon-head enterprises have established close cooperative relations with government arms and agencies, universities and other research institutes, as well as with players in the agricultural materials, food manufacturing, and food distribution

L. Kiminami (✉)
Institute of Science and Technology, Niigata University, Niigata, Niigata, Japan
e-mail: kiminami@agr.niigata-u.ac.jp

A. Kiminami
Department of Agricultural and Resource Economics, Graduate School of Agricultural and Life Sciences, The University of Tokyo, Tokyo, Japan
e-mail: akira@mail.ecc.u-tokyo.ac.jp

© Springer Japan 2016
L. Kiminami, T. Nakamura (eds.), *Food Security and Industrial Clustering in Northeast Asia*, New Frontiers in Regional Science: Asian Perspectives 6, DOI 10.1007/978-4-431-55282-6_10

industries. This has allowed them to operate highly diverse businesses, invest resources into research and development, drive innovation, and thereby gain competitive advantages. Furthermore, the advent of dragon-head enterprises has meant a new and ongoing accumulation of human, material, and financial resources, and not only from rural communities and surrounding areas, but from overseas as well. This phenomenon is the very essence of the formation of clusters, which have been defined as "geographic concentrations of interconnected companies, specialized suppliers, service providers, firms in related industries, and associated institutions (for example universities, standards agencies, and trade associations) in particular fields that compete but also co-operate" (Porter 1998). In the same vein, agricultural and rural policies that seek to create "agriculture development zones" and attract leading enterprises can be seen as cluster policies or cluster initiatives of a sort.

The purpose of this study is to assess the potential of clustering in the development of agriculture and rural communities in China. We shall examine the food industry in detail, which is the link in the food chain that propels the industrialization of agriculture, and identify instances of industrial agglomeration and business collaboration.

2 Methodology Used in This Study

Porter's (1998) Diamond Model, arguably the best-known model of industrial clusters, analyzes industrial clusters through interactions among factor conditions, demand conditions, related and supporting industries, and firm strategy, structure and rivalry. Hence, a synthesis of research into the structures and formal mechanisms of industrial clusters and quantitative research is necessary. For this reason, in this study, as Fig. 10.1 shows, we look at clusters from a comprehensive standpoint, focusing chiefly on business collaborations but also taking into account the factors that define those collaborations and their relationships to economic performance. First, case studies are used to clarify issues involving cluster initiatives and business environments. This is followed by quantitative analyses of economic effects. Two analytical methods are used. The first applies a method for analyzing industrial agglomerations to rural economies. The second method consists of quantitatively

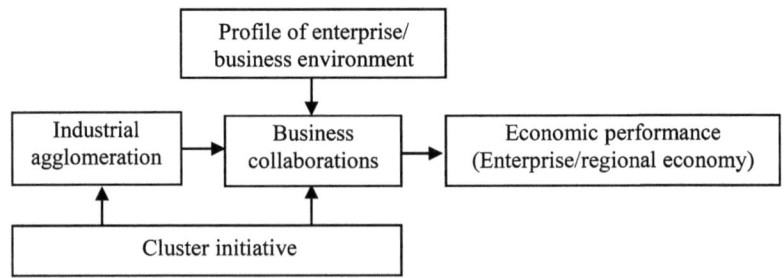

Fig. 10.1 Analytical framework in this research

analyzing the relationship between business collaborations among companies and economic performance. Collaborations and resulting economic performance levels were ascertained using questionnaire surveys targeting the food industry, the industry which drives the formation of agricultural clusters. The relationship between the two is derived from the surveys.

3 Analysis of Agricultural Clusters

3.1 Overview of the Agricultural Clusters

It is quite difficult to get an exhaustive picture of agricultural clusters in China, but we have so far conducted several surveys of representative agricultural clusters in that country (Kiminami and Kiminami 2009b). The present discussion will focus on a dairy cluster in Inner Mongolia and a hops cluster in Xinjiang.

3.1.1 The Dairy Cluster in Inner Mongolia

A dairy cluster has formed in Inner Mongolia, led by dairy companies.[1] There are 110 dairy companies in Inner Mongolia, and almost all of them have contracts with dairy farmers for procuring raw milk. In 1994 the Yili Group began construction on a dairy resource base. At present it has five large resource bases, located in Hohhot and the Hulun Buir Prairie in Inner Mongolia, and the regions of Dorbod in Heilongjiang Province, Xi'an, and Jin–Jing–Tang, where they have 300,000 head of dairy cattle and produce 300,000 tonnes of milk annually. Mengniu Dairy is a private company established in 1997. Mengniu has dairy resource bases spread out around the country, in Inner Mongolia, Heilongjiang Province, Hebei Province, Henan Province, Shanxi Province, and Qinghai Province. It produces 1,500,000 tonnes of milk annually, and has certifications in standards such as ISO 9001, ISO 14001, OHSAS 18001, CMP and HACCP.

3.1.2 The Hops Cluster in Xinjiang

The Xinjiang Uyghur Autonomous Region is known for its production of hops, and is in fact the largest hops-growing area in China. At the center of Xinjiang's hops production and processing operations and the core of the hops cluster is Xinjiang Sapporo Agricultural Science & Technology Development Co., Ltd. (hereinafter "Xinjiang Sapporo"), located in Fubei region. Xinjiang Sapporo is a joint venture which was created in 1987 through investments by the Suntime International Economic Cooperation (Group) Co. Ltd. (contributing 50 %), Sapporo Breweries

---

[1] Dagula and Kiminami (2009) offered a comparison of corporate-led dairy farming, dairy farming relying on government assistance, and traditional dairy farming.

(45 %), and Toyota Tsusho (5 %). Of these three companies, Suntime is under the administration of the 222nd Regiment of the Xinjiang Production and Construction Corps (XPCC). The Japanese side supplied the cash and the Chinese side supplied the land. The funds invested amounted to 6 million yuan, and from 1987 the registered capital of the company has risen to over 40 million yuan.

Hops production began on a trial basis in 1987, and was expanded in 1992. Production levels peaked in 2002. A subsequent supply-side surplus drove prices down, so from 2002 to 2004 production levels were cut, but since 2005 they have been on the rise again.

Tracts of 800×800 m are the standard size for hops fields operated by Xinjiang Sapporo, which are in two main locations. One is a 3,500 *mu* field where hops destined for Europe are grown. The second is a 1,600 to 2,000 *mu* field, of which the top third of the yield in terms of quality is exported to Japan. Pesticide levels for hops to be exported to Germany and the United States comply respectively with EU and U.S. standards, while hops to be exported to Japan are grown without pesticides. Experiments on different hops varieties are conducted using varieties developed exclusively by Sapporo Breweries. The development of hops varieties ordinarily takes about 20 years. Once the hops are harvested, they are either dried, pulverized and made into pellets, or made into a paste-like hops essence by extracting the bittering ingredients and oils. The hops processing procedures were certified as compliant with ISO 9002 in 1998, and with ISO 14001 and HACCP standards in 2006. Sapporo Breweries is the only importer in Japan, but there are several buyers in the Chinese domestic market, some of which have Japanese investors not related to Sapporo Breweries.

3.2 Business Collaboration

Business collaboration in the dairy cluster involves cooperation between production businesses and processing businesses. The Yili Group and Mengniu Dairy both employ the same business model, which we might call the "company–farmer" model. In this company–farmer model, the companies contract with farmers in the community. The companies provide technologies, services, and capital to the farmers, who in turn supply the companies with milk. Dairy cattle are raised in stalls and given feed three to four times a day. The feed used is mainly silage, corn stalks and feed blends. The companies give technical guidance to dairy farmers according to their contracts. The scale of these farming operations, however, is quite small, averaging about five head of dairy cattle per household.

Business collaboration in the hops cluster also takes the form of cooperation between producers and processors, but it is a cooperation among three "parties": the beer industry, the XPCC, and farmers. Most hops production is contracted out to farmers. In fact, only about 5 % of all farming land is managed directly by the company. The farms themselves, which are leased out by the XPCC, were originally used to grow wheat and rapeseed. Contracts are formed on a 12-*mu*-per-farmer basis. If there are more farmers in a farming family that can invest more labor, the contracted area increases. Technical assistance is provided by the company, and farmers work according to standards prepared by the company. Whenever farmers choose to borrow funds, the company serves as a loan guarantor. Farmers usually assume the costs of production under the contracting methods employed. However, the building of relationships between these two very different entities, i.e., a company and farmers, is said to be very difficult. One solution that could be effective in overcoming this difficulty is to create a cooperative that acts as an intermediary and deals with issues between the company and the farmers.

In both cases described above, business collaboration is largely in the form of the procurement of raw materials by food producers, and is therefore characterized by little collaboration with parties in different industries or operating different types of businesses. Generally speaking, agricultural clusters are dominated by regionally specialized clusters of companies in the same line of business. Such clusters are highly efficient and possess much growth potential, but at the same time they face problems in the area of new product development. In terms of technological development, priority tends to be placed on the improvement and stable use of existing technologies.

3.3 The Economic Performance of the Studied Clusters

Both the dairy cluster and the hops cluster have grown. Each of them has produced considerable economic results.

In the dairy cluster, the yearly net returns for farmers per milking cow are over 1,300 yuan. Silage and corn stalks are purchased from non-dairy farmers, including those in poor households. The sale of corn stalks and other materials for feed therefore also contributes to higher incomes among poor households.

In the hops cluster, company sales in 2005 amounted to 37 million yuan, and operating profits were 14 million yuan. The cluster has gained the reputation of being one of the most profitable industrial agriculture operations. Hops farmers, for their part, earn 3,000 yuan per *mu*. This is a high level of profitability for both parties, attesting to the enormous economic benefits derived from the cluster. It should also be noted that land devoted to hops production does not account for a very large ratio in terms of area within the community's agricultural economy. The 222nd Regiment of the XPCC manages 9,380 ha of farmland, of which only 470 to 540 ha is used for growing hops. Hops, however, account for 10 % of the overall income of the area farmers, making it the most economical and stable crop among those grown by the XPCC.

3.4 External Diseconomies of Agglomerations

Industrial agglomeration can cause not only external economies of scale, but also external diseconomies as well. In the dairy cluster taken up here, for example, disposal of manure from dairy cows has become an increasingly pressing problem with the increase of contracted farmers. Many farmers either use dung from dairy cows for fuel and fertilizer, or they discard it as waste. The large volume of discarded dung poses environmental and health risks, specifically contamination by and spread of bacteria. In 2006, in an effort to prevent environmental contamination and to put the manure to use, the Yili Group began biologically reprocessing manure from livestock areas into methane.

3.5 Clusters and CSR

Core companies in clusters often have corporate social responsibility campaigns. In the Inner Mongolia dairy cluster, these companies have poverty assistance programs for impoverished regions and the poor in general. Unlike measures to fight poverty undertaken by the government, the poverty assistance programs of the Yili Group and Mengniu Dairy are comparatively small in their respective scopes. That said, most of the assistance provided through these anti-poverty activities is gratis. Assistance from the companies has become invaluable to areas not covered by analogous government programs, poor students, those with disabilities, and other disadvantaged people.

Also, in the hops cluster, the XPCC has constructed housing for farmers. In addition to providing subsidies for construction costs, the XPCC also does construction work for essential services such as running water, etc.

3.6 Clusters and Market Competitiveness

The respective competitive environments of the agricultural clusters dealt with in this study differ in that the dairy cluster is relatively competitive, while the hops cluster is relatively monopolistic. In the diary cluster, intense competition over dairy resource bases has given rise to large discrepancies in the purchase price of raw milk among the companies. This has led to the problem of farmers selling milk to a company with which it does not have a contract out of a desire to maximize their selling prices. Furthermore, the phenomenon of "over-commitment" on the part of farmers has become a problem. For example, in some areas, there are two milking facilities owned by two different companies within the same village and the total demand for milk by the two companies ends up exceeding the production capacity of that village due to conflicting commitments. The ultimate cause is the fierce competition between the two companies over the milk supply, yet the result is that the profits of the companies and farmers alike are destabilized, and business sustainability suffers on the part of both the farmers and the companies. For this very reason, alternative business models that would stabilize the supply of raw milk resources are currently being explored.

These phenomena suggest that there is no unique relationship between market competitiveness and cluster development. They imply that cluster development is facilitated by competitive markets in some cases, but facilitated by monopolistic markets in others. A further observation is that the dairy cluster appears to conform to the external economies hypothesized by Porter, while the hops cluster corresponds to the external economies of Marshall, Arrow, and Romer.

3.7 Clusters and Regional Characteristics

Both the dairy cluster and the hops cluster are largely dependent upon the natural conditions specific to the respective areas. This is a major feature of agricultural clusters. For this reason, the cluster models discussed here are not necessarily reproducible in a generalized way. Instead, in the agricultural sector, clusters must be created in a manner that makes the most out of the advantageous features of each region.

3.8 Cluster Initiatives

Table 10.1 shows a comparison of the dairy cluster and the hops cluster analyzed by cluster initiative models.

3.9 Demand Conditions for Cluster Formation

Porter's Diamond Model proposes demand as a condition for clustering and hypothesizes a relationship in which the pressure of customers with high levels of demand drives corporate innovation. Demand conditions have an effect on the transition from low-quality products and services that are easy to imitate to competitiveness rooted in differentiation, and with the advent of globalization the demand within a particular region dictates a shift in importance from large-scale volume to quality. It should be mentioned here that, generally speaking, once per capita income increases as a result of economic growth, food consumption does increase, but with that increase there is an accompanying trend in which consumption of relatively higher-quality goods increases, while consumption of lower-quality goods decreases. In addition, once a certain income level is exceeded, the quality in food consumption becomes more important than quantity. One can see evidence of this changing pattern of food consumption in China as well, as the country continues on its path of economic growth. In Kiminami and Kiminami (2009a), the authors used the example of rice to conduct a survey by questionnaire in Shanghai targeting consumers of different income levels in order to determine which factors consumers base their choices on when purchasing food products. The six criteria presented as choices in the questionnaire for the purchase of rice were: price; taste; brand name; place grown; milling date; and cultivation method.

Table 10.1 Summary of cluster initiatives

Sector		Dairy farming	Hops
Region		Inner Mongolia	Xinjiang
Setting	Production base	Within province	Within city
	Level of regional economic development	Low	Low
	Capital of core-organization	Domestic corporation	Joint venture corporation
	Products market	Domestic	Foreign > domestic
	Demand	Expansion	High quality, differentiation
	Functions of research and development	Domestic	Foreign
	Related policy	Model project of technology research and industrialization on dairy industry	Basic policy on economic development of "The 11th 5 year Plan for Xinjiang Uygur Autonomous Region Economic Development"
		Poverty reduction policy	
Objectives	Research and networking	Milk association	Joint venture corporation (Japanese corporation)
	Commercial cooperation	Solving environment problems	Promotion of regional economy
	Innovation and technology	Stable procurement of raw materials	Stable production technology of high quality hops
	Cluster expansion	Procurement of raw materials	Foreign capital introduction
Process	Initiation and planning	Corporation	Xinjiang production and construction corps + foreign capital
	Governance and finance	Corporation	Xinjiang production and construction corps + foreign capital
	Framework	Provincial level	City level
Performance	Innovation	Organization	Process
	Improvement of competitiveness	Domestic	International (export to Japan, U.S. and Europe)
	Cluster growth	New farm formation	Increased employment
	Goal fulfillment	Growth of farmers' income	Growth of farmers' income improvement of living

4 The Economic Effects of Industrial Agglomeration

It is difficult to clarify the economic effects of agricultural clusters statistically. Kiminami and Kiminami (2009b) analyzed the effects of industrial clustering on rural development in China through the estimation of a production function.

The index of agglomeration of TVEs (*IA*) is defined as follows:

$$IA = NTVE / NTV \tag{10.1}$$

where, *NTVE* is the number of Township and Village Enterprises, and *NTV* is the number of towns and villages.

The Cobb-Douglas production function of TVEs is estimated by using the data for 31 provinces from China Agricultural Statistics 2005:

$$\log V = a_1 + a_2 \log L + a_3 \log K + a_4 \log IA \tag{10.2}$$

where, *V* is the average value added of TVEs (10,000 yuan), *L* is the average number of employees of TVEs (persons), and *K* is the average net value of the fixed assets of TVEs (10,000 yuan).

The result of estimation (10.2) is as follows:

$$\log V = -0.480 + 0.743 \log L + 0.533 \log K + 0.174 \log IA$$
$$(2.70)* \quad (2.97)* \quad (3.45)* \quad (3.18)*$$
$$\text{adj. } R^2 = 0.947$$

where the numbers in parentheses are the *t*-values, and "*" denotes statistical significance at a 1 % level.

The result of the above estimation shows that the parameter *IA* is positive and statistically significant. It means that the higher rate of agglomeration of TVEs brings higher value added for TVEs, which makes it clear that the agglomeration improves the economic performance of TVEs.

Second, we shall clarify the effect of the growth of TVEs on farmer income by estimation using the formula as below:

$$\log Y = b_1 + b_2 \log(TV / N) \tag{10.3}$$

where *Y* is the per capita annual income of farmers (yuan), *N* is the population of rural areas (10,000 persons), and *TV* is the total value added for TVEs in each province (10,000 yuan).

The result of estimation (10.3) using the data for 30 provinces[2] from China Agricultural Statistics 2005 is as follows:

$$\log Y = 2.067 + 0.392 \log(TV / N)$$
$$(12.93)*(9.09)*$$
$$\text{adj.} R^2 = 0.738$$

where the numbers in parentheses are the *t*-values, and "*" denotes statistical significance at a 1 % level.

---

[2] Tibet is excluded from the analysis because of the lack of farmer income data.

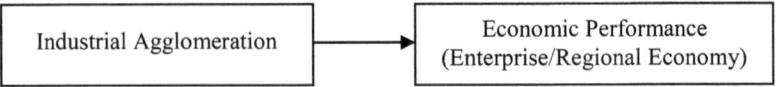

Fig. 10.2 The relationship between the industrial agglomeration and economic performance indices

The result of the above estimation shows that the parameter *TV/N* is positive and statistically significant. It means that the higher per capita value added of TVEs brings higher farmer income which makes it clear that the growth of TVEs increases farmer income.

Figure 10.2 shows the relationship between the industrial agglomeration and economic performance indices in the analysis above. The result shows that the agglomeration of enterprises in rural areas improves the performance of enterprises as well as increasing farmers' income. Therefore, it is considered that the strategies of industrialization through an agglomeration economy are effective for the rural economy.

5 Conclusions

The results of the analyses of agricultural clusters in China in this study suggest that clusters do achieve certain positive results in the way of development in the agricultural sector and rural communities. We believe that part of this is attributable to the fact that policies of agricultural industrialization, the creation of agriculture development zones, and other such policies in China share much in common with the country's industrial cluster policies. Furthermore, utilizing the potential benefits with regard to the development of regional economies through external economies of industrial agglomeration is promising as a method for rural development. However, mere industrial agglomerations are not enough to sustain regional economic growth. There must also be cluster initiatives in place to facilitate clustering, as well as industrial development policies that promote business collaborations among different companies.

China's economy continues to grow at a high rate. With the higher income levels and broader income gaps that come with this growth, food demand is both increasing in volume and diversifying. As a result, there is stronger competition in the food sector, competition that is rooted in product diversification. This development represents an ongoing transition in China's food market from a period of expansion to a period of maturity. On the one hand, this means that a broad range of demand conditions for cluster formation will be met. Therefore, this transition has the potential to facilitate cluster formation. On the other hand, however, it also means that a shift from intra-industry to inter-industry business collaboration will be necessary for the sustained development of those clusters.

An earlier version of this paper was presented at the International Association of Agricultural Economists Conference, Beijing, China, 16–22 August 2009.

References

Dagula, Kiminami L (2009) Promotion of dairy farming and poverty reduction in Inner Mongolia, China. China Agric Econ Rev 1(1):82–96

Kiminami L, Kiminami A (2009a) Economic growth and food policy in urban China. J Chinese Econ Foreign Trade Stud 2(1):18–30

Kiminami L, Kiminami A (2009b) Rural development through industrial clustering: a case study from China. China–USA Business Rev 8(1):25–33

Porter M (1998) Clusters and competition: new agendas for companies, governments, institutions. On competition. Harvard Business School Press, Boston, pp 197–287

Wang Y (2011) The research on the effect of farmers employment creation based on agriculture industry cluster. In Zhou M (ed) ISAEBD 2011, Part I, CCIS 208. pp 184–190

Chapter 11
Industrial Agglomeration of the Food Industry in China: An Analysis of Data by Province

Hironori Yagi

Abstract On the basis of statistical data from provinces and autonomous regions, we analyze trends in industrial agglomeration in the food industry using specialization coefficients and the level of gross industrial output, and examine their relationships with the rate of return in the industry. We showed that the level of gross industrial output of China's three major food industries (food processing, food manufacturing, and beverage manufacturing) which accounts for 7.5 % of all industries, increased 1.66 times in the period between 1999 and 2003. Agglomeration had been even accelerated in provinces with high rate of agglomeration at the beginning of the period. Provinces with a large share of food processing and manufacturing production had a growing share of the output of food industry as a whole; however, those with a large share of beverage manufacturing production did not necessarily account for a large portion of food industry production as a whole. The positive correlation between the specialization coefficient and the profit margin on sales in food industries supports the positive externality effects of agglomeration within the industry.

Keywords Industrial agglomeration • Food industry • Provincial statistics • Specialization coefficients

1 Purpose and Analytic Method of This Chapter

According to the Industrial Classification for National Economic Activities (GB4754-84), China's food industry is classified into small, medium, and large categories. The "large" category is further divided into four types of industries: food processing, food manufacturing, beverage manufacturing, and tobacco processing. Regarding the first two types, this paper compares raw materials and finished goods

---

H. Yagi (✉)
Department of Agricultural and Resource Economics, Graduate School of Agricultural and Life Sciences, The University of Tokyo, Tokyo, Japan
e-mail: ayouken@mail.ecc.u-tokyo.ac.jp

© Springer Japan 2016
L. Kiminami, T. Nakamura (eds.), *Food Security and Industrial Clustering in Northeast Asia*, New Frontiers in Regional Science: Asian Perspectives 6, DOI 10.1007/978-4-431-55282-6_11

and classifies industries that produce goods with a higher degree of processing, such as food manufacturing (Shiraishi 2000). Hereafter, three of the four broad categories of industries, excluding tobacco processing, are analyzed.

In this section, we present a discussion on the relationship between food industry agglomeration and profitability by investigating the direction of future investment and development in China's food industry. We aim to contribute to food security in Northeast Asia by gathering basic information for the promotion of development and economic cooperation in this region's food industry. However, conducting an analysis using a more detailed industry classification or examining agglomeration trends or causes within each province is beyond the scope of this study.

In general, an industrial agglomeration exists in three forms: (i) large-scale expansion of a company using the same management, (ii) regional concentration of several management teams in the same types of industries; and (iii) urbanization due to various management teams in different types of industries (Hoover 1937). The respective advantages that accompany agglomeration ("agglomeration economies") are caused by the following: (i) economies of scale or internal economies (internal from the perspective of management), (ii) economies of regional agglomeration or external economies; and (iii) economies of urbanization (external for both management and industry) (Nishioka 1976). In particular, the fact that the external effects of the concentration of similar and different types of industries are important factors in industrial agglomeration, as Marshall (1890) pointed out, is receiving renewed recognition as "industrial clusters," even in the context of the highly networked modern economy (Porter 1998).

The Industry Location Society (1967) once ordered the industrial agglomeration of food manufacturing by trends in the location of each industry (Table 11.1). Ueji and Kajikawa (2004) ordered it by the concentration of identical management in the food industry, that is, by oligopoly. Glaeser et al. (1992) suggest that the degree of concentration in relevant industries and the degree of local monopolization, competitiveness, diversity, and initial conditions influence the increase in employment in specific industries in cities. Here we can use the following specialization coefficient to measure the degree of concentration in industry. In other words, s_{ij}, the specialization coefficient for industry j in region i, can be expressed as follows:

$$s_{ij} = p_{ij} / \overline{p}_j \tag{11.1}$$

where p_{ij} is the share of industry j in region i and $\overline{p}_j$ is the share industry j has of the national average.

Below, we calculate the specialization coefficients of the food industry, within China's provinces and autonomous regions, and across broad categories, to determine whether changes in these coefficients are relevant to geographic trends and industry profitability. We decided to analyze the years 1999 and 2003, a 4-year time frame sufficient to compare the types of industries. First, in Sect. 2, we outline trends in the level of gross industrial output across provinces and autonomous regions, as well as across broad categories of the food industry. Next, in Sect. 3, we

Table 11.1 Relationship between urban concentration and location of food production

Relationship to urban concentration				Other factors with a greater relationship		
Metropolis	Suburban areas	Provincial cities	Countrysides	Specialized product industries	Industry dependent on local resources	Related equipment industries
Sauce (other than soy sauce)	Other seasonings	Soft drinks	Grape sugar (glucose)	Cereal milling	Meat products	Monosodium glutamate
Sugar	Baking powder, yeast	Beer	Ice-making	Pastries	Daily products	
Other oils and fats for cooking				Noodles	Marine products	
				Tea	Canned vegetables	
				Malt	Plum wine	
				Unclassifiable foods	Sake	
					Distilled liquor	
					Starch	

Source: Excerpted from Industrial Location Society (1967)

infer trends in food industry agglomeration from trends in the specialization coefficient. Finally, in Sect. 4, we analyze the relationship between the specialization coefficient and profitability.

2 Trends in the Food Industry in China

Table 11.2 shows the level of gross industrial output across three broad categories of the food industry in 1999 and 2003. It should be noted that the level of gross industrial output is equivalent to the gross national product plus intermediate investments. Reflecting China's rapid industrial growth, in these 4 years, the food processing industry grew 1.7 times, the food manufacturing industry 1.8 times, and the beverage industry 1.3 times.

If we look at the level of the gross industrial output of the food industry (Table 11.3), the three industries account for 7.5 % (2003), a slight decline from 1999.

Next, Figs. 11.1 and 11.2 show, on a quantitative basis, the changes in the output of specific intermediate commodities through real figures and percentage change (1999 = 100 %). Only rice output decreased; output of all other types of food increased, especially canned goods, vegetable oil for cooking, and dairy products, which increased by more than twice the 1999 amount.

In addition, Fig. 11.3 shows the level of gross industrial output of the provinces and autonomous regions, in 1999 and 2003, over the three broad food industry categories. The higher the industry output is above the 45° line in the graph, the greater is the increased production over the 4-year period. In the food processing industry, Shandong Province has a market share of over one-third of the total, unrivaled by

Table 11.2 Changes in the level of gross industrial output (Unit: Hundred million yuan)

	1999	2003
Processed foods industry	3,517.0	6,152.3
Food manufacturing industry	1,262.2	2,290.1
Beverage manufacturing industry	1,658.7	2,233.2

Source: Sea Press (2005), China's Food Industry 2006

Table 11.3 Changes in the level of gross industrial output as a percentage of all industries

	Share in 1999	Share in 2003
Processed foods industry	4.84 %	4.32 %
Food manufacturing industry	1.74 %	1.61 %
Beverage manufacturing industry	2.28 %	1.57 %

Source: Calculated using the level of gross industrial output listed in the 1999 and 2003 editions of the China Statistical Yearbook

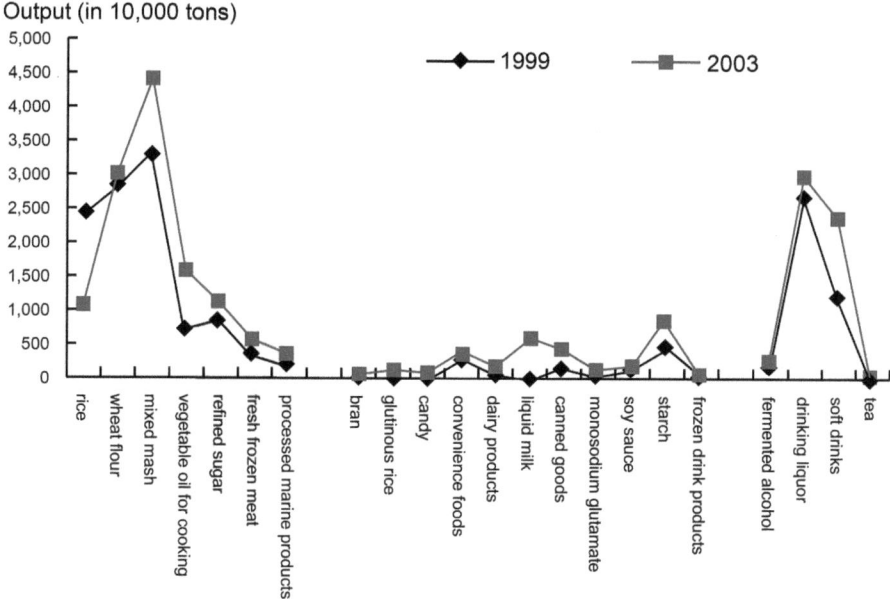

Fig. 11.1 Changes in food industry output by product (1999–2003). Source: China Food Industry Yearbook

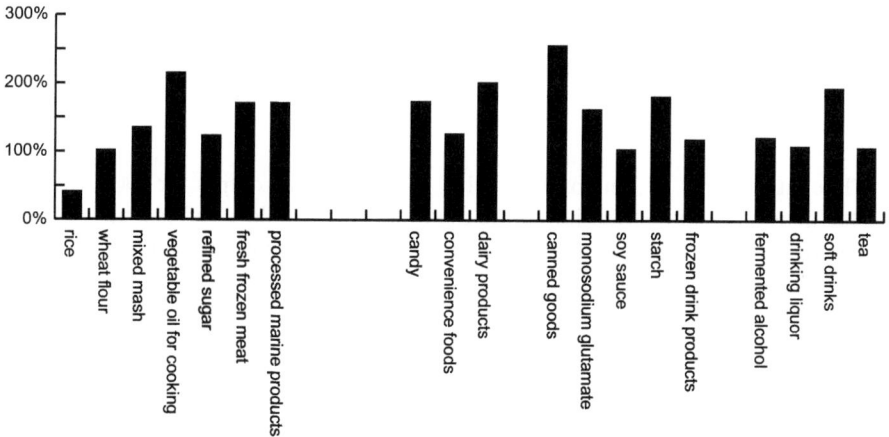

Fig. 11.2 Changes in food industry output by product [2003 compared to 1999 (base year)]. Source: China Food Industry Yearbook

other provinces. Moreover, Shandong has overtaken Guangdong Province and has become the leader in the food manufacturing industry. On the other hand, in the beverage industry, while Shandong and Guangdong's market share is large, in recent years Sichuan Province has seen remarkable growth.

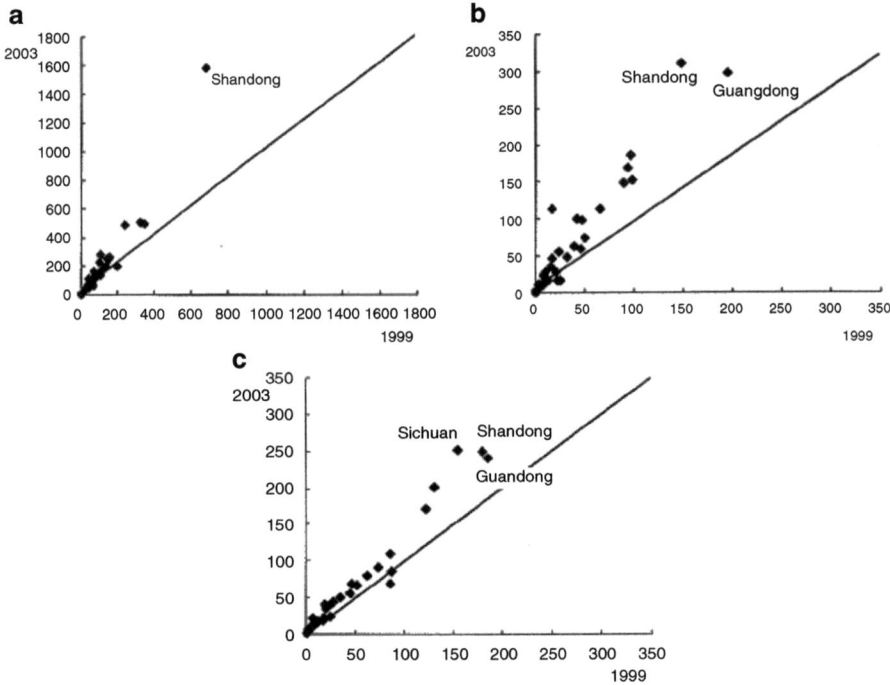

Fig. 11.3 (**a**) Changes in the level of gross industrial output for each province and autonomous region (food processing industry), (**b**) Changes in the level of gross industrial output for each province and autonomous region (food manufacturing industry, (**c**) Changes in the level of gross industrial output for each province and autonomous region (beverage manufacturing industry) (*Source*: Compiled from the China Food Industry Yearbook (**a** through **c**))

To continue, Fig. 11.4 maps the level of gross industrial output for 2003, showing us the geographical distribution of the food industry. The level of production in the food processing industry is high in Shandong and its environs, the Guangxi Zhuang Autonomous Region, coastal areas such as Guangdong, and inland areas such as Sichuan. In the food manufacturing industry, in addition to Shandong and Guangdong, the level of production has increased even in northern areas such as the Inner Mongolia Autonomous Region and Heilongjiang Province. In the beverage manufacturing industry, we see that the market share is high in Sichuan and coastal areas such as Shandong and Guangdong.

3 Trends in Food Industry Agglomeration by Province

Here, we look at industrial agglomeration trends in the food industry, by province and autonomous region, via the specialization coefficients for the level of gross industrial output. By looking at the specialization coefficients rather than the level

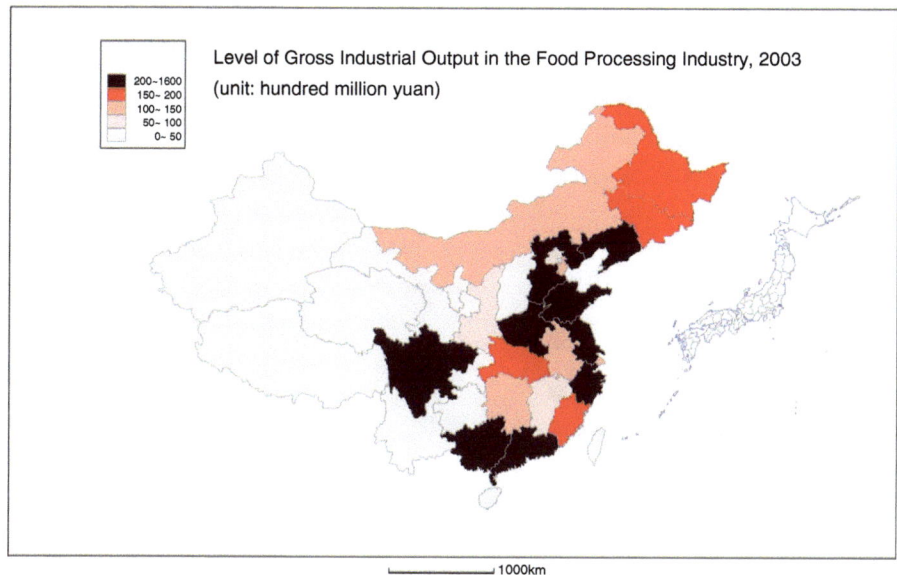

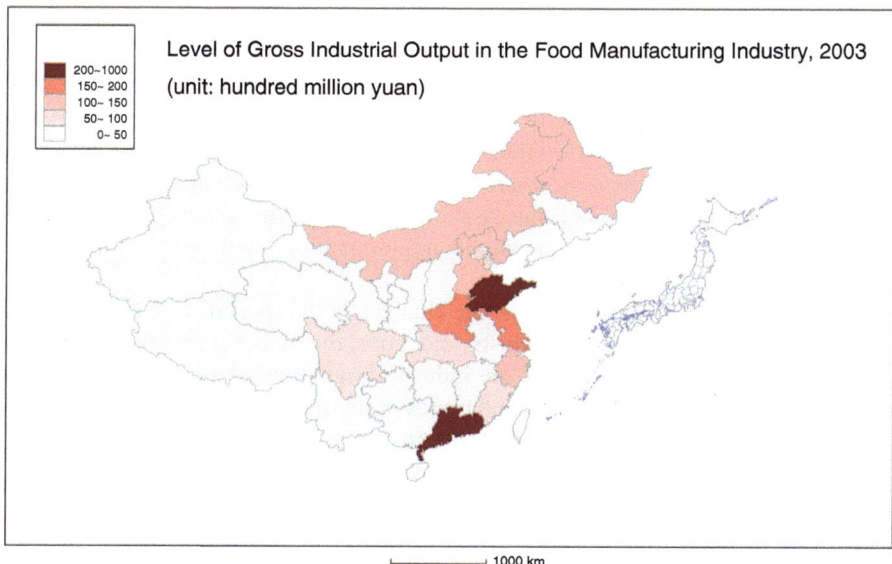

Fig. 11.4 Geographical distributions of the level of gross industrial output in the food industry (2003)

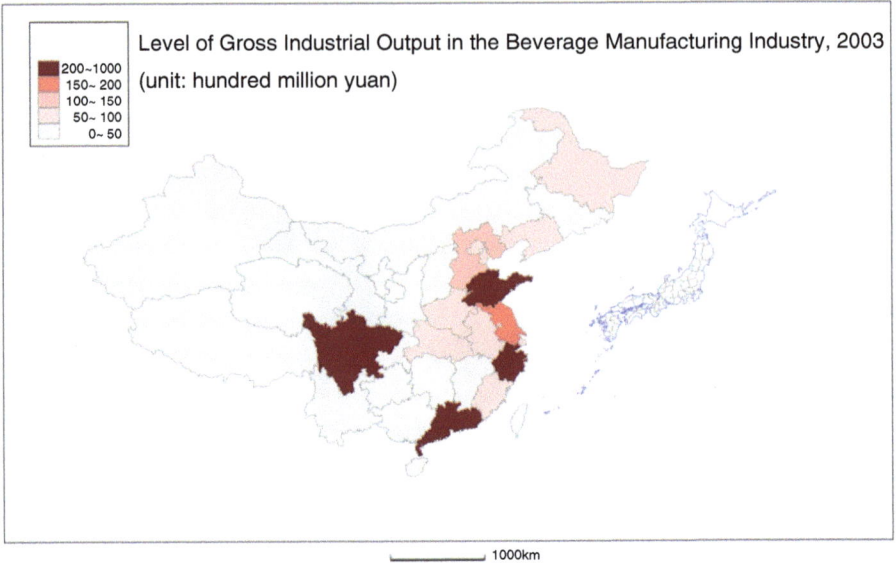

Fig. 11.4 (continued)

of output, we can determine the degree of agglomeration in an industry relative to other industries in the region. In other words, the computation of the specialization coefficients shows that regions where industrialization came early took a greater amount of production to become "specialized"; on the other hand, in late-industrializing regions, even a slight agglomeration will make the region "specialized."

First, the graphs in Fig. 11.5 show the changes in the specialization coefficients for 1999 and 2003 in each food industry category; the higher a data point is above the 45° line, the more advanced is the industrial agglomeration. Also, the supplementary tables (a–c) show changes and trends in the specialization coefficient, relative to the specialization coefficient for 1999 (1 or more, 1 or less). The larger the value in the shaded portions of the table, the more the advancement of agglomeration in higher-ranked regions, and the more the decline of agglomeration in lower-ranked regions.

In the food processing industry, Shandong's share of gross industrial output level appeared dominant, but if we look at the specialization coefficient, the Guangxi Zhuang Autonomous Region was leading in all years; moreover, its specialization was advancing. Overall, the value of the correlation coefficient over the 2 years was a high of 0.93; in addition, as shown in the supplementary chart, regions with a large specialization coefficient (1 or more) in 1999, showed an increase in their specialization coefficient and concentration.

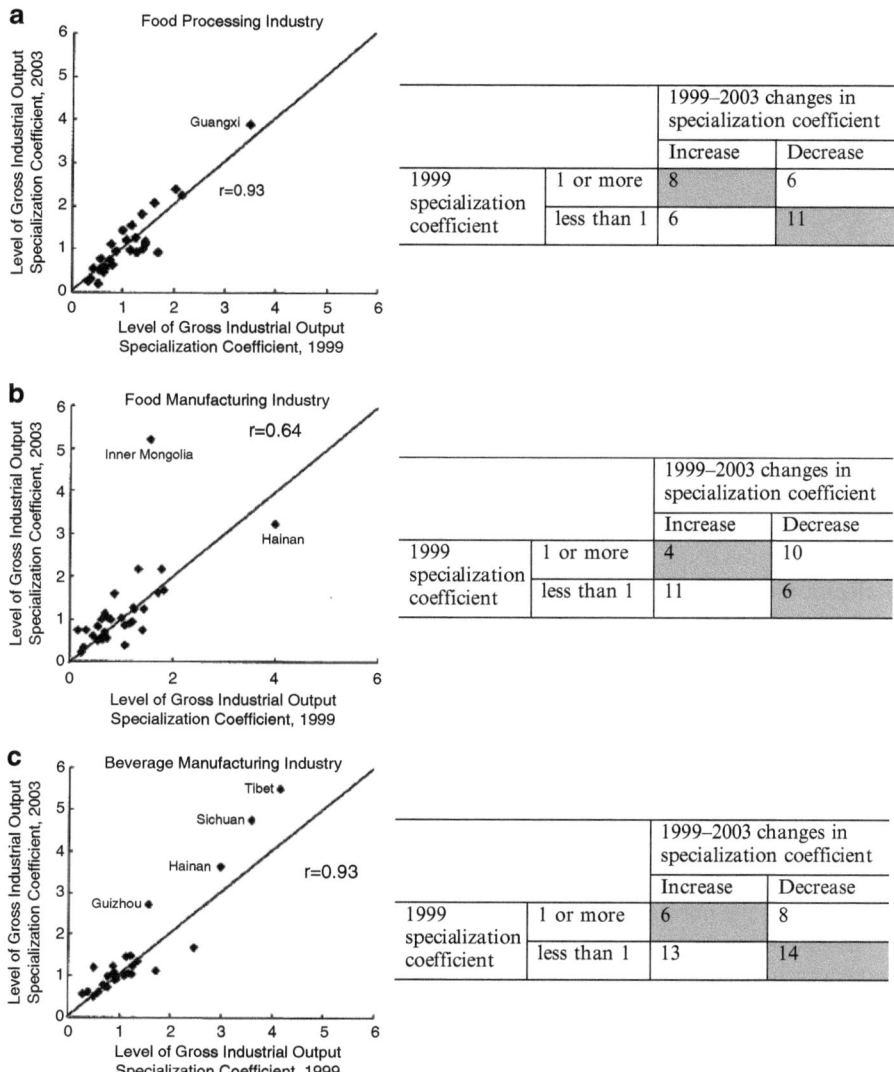

Fig. 11.5 Changes in specialization coefficient by province and autonomous region (using the level of gross industrial output in both 1999 and 2003 as bases). (**a**) Food processing industry, (**b**) food manufacturing industry, (**c**) beverage manufacturing industry

In the food manufacturing industry, we saw a decrease in the specialization coefficient for Hainan Province, which was top in the industry in 1999, and a rapid increase in industrial agglomeration in Inner Mongolia. The correlation coefficient for the 2 years is low (0.64); even in the supplementary table, we see a decrease in the specialization coefficient for regions with agglomerations in 1999, and an increase for regions where there was originally no agglomeration. In summary, in food manufacturing, the specialization coefficients have seen drastic up-and-down fluctuations.

Table 11.4 Changes in the standard deviation of the specialization coefficient over province and autonomous region

	1999	2003
Food processing industry	0.66	0.77
Food manufacturing industry	0.72	0.98
Beverage manufacturing industry	0.90	1.18

Source: Compiled from the China Food Industry Yearbook, n=31

In the beverage manufacturing industry, the top three regions of Tibet, Sichuan, and Hainan had increasing specialization coefficients, and industrial agglomeration was progressing. Although the correlation coefficient is high for the 2 years under study (0.93), if we take the specialization coefficient for 1999 as standard (whether it is 1 or greater or less than 1), as in supplementary table (c), we see that the agglomeration trend is not necessarily stable.

To determine the differences in industrial agglomeration among regions, we calculated the standard deviation of the specialization coefficient (Table 11.4). The figures increase across all broad food industry categories, indicating an increasing degree of regional specialization. The beverage manufacturing industry had the specialization coefficient with the greatest standard deviation, followed by the food manufacturing and food processing industries.

Next, Fig. 11.6 shows the geographical distributions of the specialization coefficients for the level of gross industrial output in 2003. In the food processing industry, we see agglomerations in Guangxi, Shandong, and Henan Provinces. In the food manufacturing industry, we see agglomerations in the northern parts of Hainan, Henan, and the interior. The beverage industry is characterized by agglomerations in the southern part of the interior and Hainan.

To determine the type of products that comprise the above industrial agglomerations, we extracted the production output rankings (2003) for intermediate goods classified by the main provinces and autonomous regions (Table 11.5).

In Shandong, the leading producer in the food processing and manufacturing industries, rice, frozen drinks, and tea are low-ranked; however, other goods including those in the beverage manufacturing industry occupy the top ranks. Further, even Guangdong, which is ranked second in food manufacturing, is in the top ranks excluding some commodities. In contrast, Sichuan, which leads the beverage manufacturing industry and has continuous agglomeration, ranks high in most products in the food processing industry, but lags slightly in the food manufacturing industry.

Therefore, if we examine the case of northeastern China – the liquid milk and frozen foods of Inner Mongolia; the dairy products, fermented spirits, and rice of Heilongjiang; the starch of Jilin Province; and the processed marine products of Liaoning Province, among others – we learn that the development of a food industry with links to raw material resources is worthy of attention. However, compared to states provinces that rank high in the food industry, such as Shandong and Guangdong, the rank of northeastern China's other commodities cannot be high, and it would appear that the links between different commodities are not completely developed.

11 Industrial Agglomeration of the Food Industry in China: An Analysis of Data by... 151

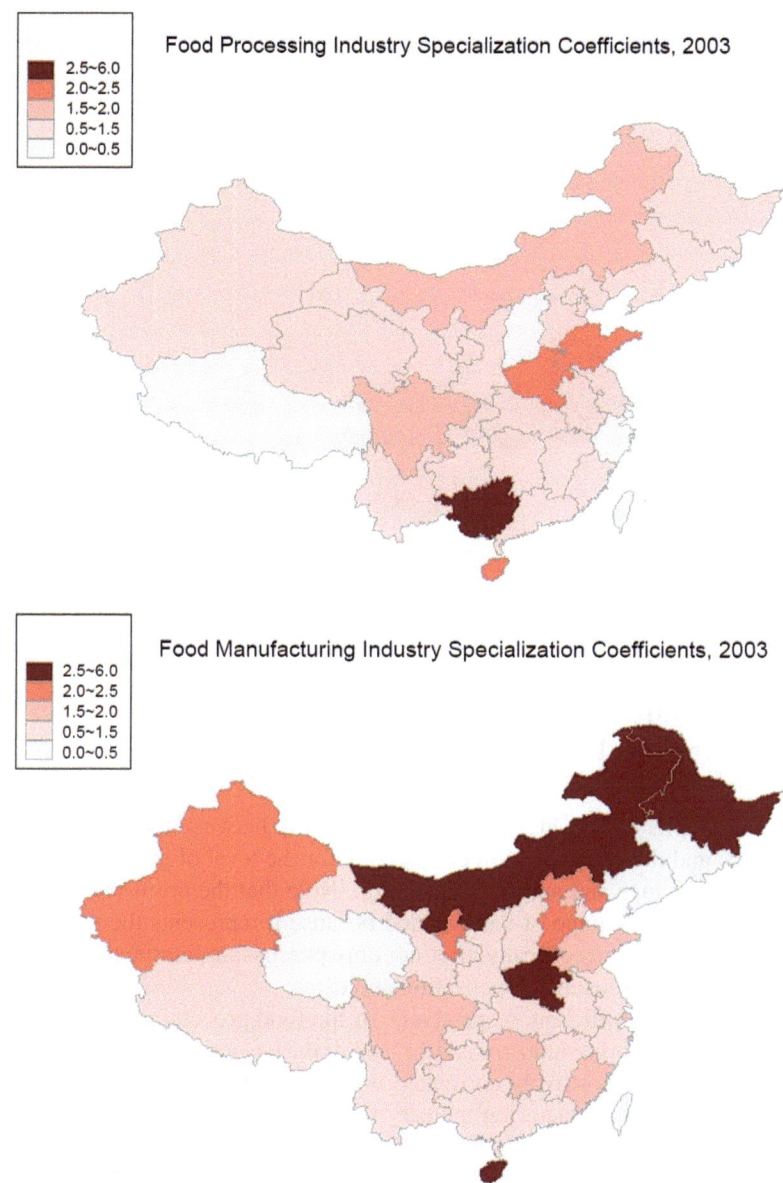

Fig. 11.6 Specialization coefficients of the level of gross industrial output, 2003

Fig. 11.6 (continued)

4 Relationship Between Industrial Clustering and Earnings Rates

Finally, the graphs in Fig. 11.7 show the relationship between the profit margin on sales (2003) and the specialization coefficients for the level of gross industrial output for each province and autonomous region. Note that the profit margin on sales differs from the percentage of value added because it represents the percentage of earnings retained by the company after not only purchasing intermediate inputs but also after paying employee compensation and taxes.

The correlation coefficient, respectively, of the food processing industry is 0.24, of the food manufacturing industry is 0.21, and of the beverage manufacturing industry is 0.67; these are positively correlated with the profit margin on sales. Among these, the beverage manufacturing industry shows a highly positive correlation, and the industry's profit margins on sales in Tibet, Guizhou, and Sichuan are increasing.

5 Conclusion

In this chapter, on the basis of statistical data from provinces and autonomous regions, we analyzed trends in industrial agglomeration in the food industry using specialization coefficients and the level of gross industrial output, and examined their relationships with the rate of return in the industry. We showed that the level of

Table 11.5 Ranking of output per commodity in each province (2003), among 31 regions

Province, autonomous region		Shandong	Guangdong	Sichuan	Tibet	Guangxi	Northeast China Inner Mongolia	Heilongjiang	Jilin	Liaoning
Remarks of each Province		Ranked #1 in production in food processing and manufacturing industries	Ranked #2 in production in food manufacturing industry	Ranked #1 in output in beverage manufacturing industry, clustering	Clustering in beverage manufacturing industry	Clustering in food processing industry	Clustering in food manufacturing industry	–	–	–
Food processing industry	Rice	23	10	8	28	17	18	3	12	4
	Wheat flour	1	7	11	16	20	19	14	30	12
	Mixed mash	2	1	3	31	6	16	22	13	8
	Vegetable oil for cooking	1	9	4	31	12	16	8	18	7
	Sugar	10	3	8	29	1	7	6	16	11
	Fresh frozen meat	1	5	2	31	21	6	11	17	4
	Processed marine products	1	5	25	28	9	18	23	19	3

(continued)

Table 11.5 (continued)

Province, autonomous region		Shandong	Guangdong	Sichuan	Tibet	Guangxi	Northeast China			
							Inner Mongolia	Heilongjiang	Jilin	Liaoning
Food manufacturing industry	Bran products	6	1	28	29	24	17	22	27	9
	Candy	1	3	10	28	18	23	21	24	12
	Glutinous rice	3	1	10	28	18	25	17	24	23
	Convenience foods	2	4	15	29	19	21	12	18	7
	Dairy product	2	13	7	25	30	5	1	27	20
	Liquid milk	5	12	22	31	24	1	4	21	10
	Canned goods	5	8	10	29	9	23	19	28	13
	MSG	2	6	12	29	15	23	24	16	7
	Soy sauce	2	1	10	31	19	25	16	18	8
	Starch	1	10	17	30	7	8	5	3	12
	Frozen drinks	17	4	12	28	15	1	19	10	6
Beverage manufacturing industry	Fermented alcohol	1	11	9	30	6	19	2	4	14
	Drinking alcohol	1	3	8	30	21	17	4	14	5
	Soft drinks	7	2	12	30	17	26	22	13	14
	Tea	17	9	8	27	11	23	19	25	24

Source: Compiled from China's Food Industry, 2006, Sea Press (2005)

11 Industrial Agglomeration of the Food Industry in China: An Analysis of Data by...   155

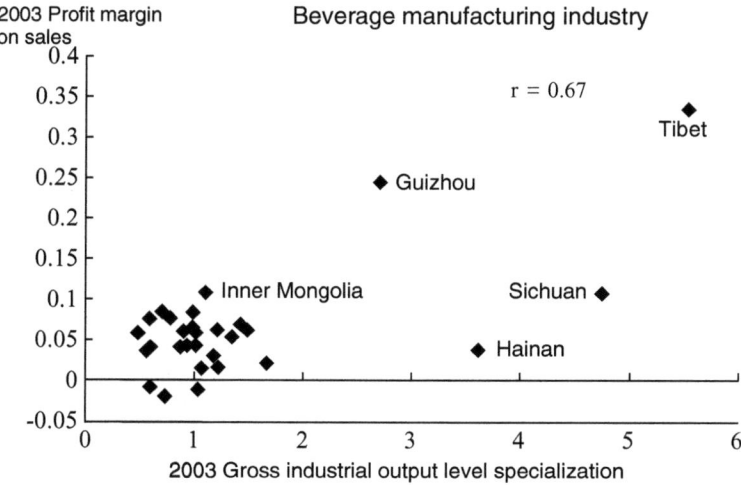

Fig. 11.7 Specialization coefficients of gross industrial output level by province and autonomous region and its relation to the profit margin on sales (2003)

gross industrial output of China's three major food industries, which accounts for 7.5 % of all industries, increased 1.66 times in the period between 1999 and 2003. Among these findings, we observed a tendency for agglomerations to progress even more in regions that had clusters from the beginning. We also showed that there is a positive correlation between the specialization coefficient and the profit margin on sales in all categories.

The provinces with a large share of food processing and manufacturing production (Shandong and Guangdong) had a growing share of the output of the food industry as a whole; however, those with a large share of beverage manufacturing production (Sichuan) did not necessarily account for a large portion of food industry production as a whole. Further, looking at the northeastern part of China, we see that, while the production of some commodities ranked high in the entire country, that of other commodities was not high. Judging from this, we find that there is a difference between (1) the clustering of the entire food industry (Shandong, Guangdong), (2) the clustering of the beverage industry (Sichuan), and (3) the clustering of some commodities (the northeastern regions).

This analysis does not sufficiently touch upon geographical clustering in the food industry because it used large, aggregate data collected from provinces and autonomous regions. This issue can be addressed, within statistical limits, by examining data collected through case studies and the like, and by considering the real state of industrial agglomeration based on food. These, however, are topics for future consideration.

This chapter is the product of research conducted by the independent enterprise of the Economic Research Institute for Northeast Asia (ERINA), undertaken in the fiscal year 2007 in "Food Security for Northeast Asia (1)". The manuscript was reproduced in the report *Food Security for Northeast Asia (1): Joint research of ERINA/Niigata University/University of Tokyo*, published in March 2008.

References

Glaeser EL, Kallal HD, Scheinkman JA, Shleifer A (1992) Growth in cities. J Polit Econ 100(6):1126–1152
Hoover EM (1937) Location theory and the shoe and leather industry. Harvard University Press, Cambridge, MA [Translated by Nishioka H (1968) Economic location theory. Taimeido Publishers, Tokyo]
Industry Location Society (1967) Study of industrial classification by regional structure and location in Japanese industry. . International Trade and Industry Research Company, Tokyo [in Japanese]
Marshall A (1890) Principles of economics. Macmillan Publishers, London
Nishioka H (1976) Analysis of economic geography. Taimeido Publishers, Tokyo [in Japanese]
Porter ME (1998) On competition. Harvard Business Press, Boston
Shiraishi K (2000) The current status and outlook of China's food industry. Rural Culture Association, Tokyo [in Japanese]
Ueji T, Kajikawa C (2004) Industrial organization theoretic study of the food industry. Agriculture and Forestry Statistics Association, Tokyo [in Japanese]

Chapter 12
The Agricultural Industrialization of China's Heilongjiang Province

Jiao Jiang

Abstract This paper refers to the fundamental distinguishing characteristics of agriculture in Heilongjiang Province, and analyses the current state of agricultural industrialization including: the regionalization of agricultural production centered on food production; the development trends which emphasize such matters as the quality of commodities; processing capacity; the cultivation of brand-name goods; and production models. In addition it raises problem points which should be paid attention to for the development of agricultural industrialized production.

Keywords Commoditized foodstuffs • Industrialization of agriculture • Production model • Agriculture in Heilongjiang Province

1 The Basic Characteristics of Agricultural Production

1.1 High Food Commoditization Rate

Heilongjiang Province is ranked top in China for area of agricultural land, and also occupies the nation's top spot for the area for food cultivation. In addition, it is also a province where the area of land able to be brought into cultivation is one of the largest.

The area planted with food crops for Heilongjiang Province made up 10.3 % of the total for China in 2008, and the total volume of production was 42.25 million tonnes (Table 12.1). The food commoditization rate was extremely high at approximately 65 %, and it is forecast that it will be on an increasing trend in the future also. In the volume of food produced which was commoditized, Heilongjiang accounted for approximately 30 % of the national total, and is in the foremost position nationally.[1]

---

[1] Based on Ministry of Agriculture of the People's Republic of China, China Agriculture Statistical Report, various issues of 2000–2008, China Agriculture Press and Heilongjiang Statistical Bureau, Heilongjiang Statistical Yearbook, various issues of 2000–2009, China Statistical Press.

J. Jiang (✉)
Heilongjiang Academy of Agricultural Sciences, Harbin, China
e-mail: hhnjj@163.com

© Springer Japan 2016
L. Kiminami, T. Nakamura (eds.), *Food Security and Industrial Clustering in Northeast Asia*, New Frontiers in Regional Science:
Asian Perspectives 6, DOI 10.1007/978-4-431-55282-6_12

Table 12.1 Volumes of foodstuffs and of commoditized food production in 2008

Ranking	Food production volume (10,000 tonnes)		Per capita production volume (kg)		Commoditized food production (10,000 tonnes)	
1	Henan	5,365.50	Heilongjiang	1,104.90	Heilongjiang	2,695.00
2	Shandong	4,260.50	Jilin	1,040.40	Jilin	1,747.80
3	Heilongjiang	4,225.00	Inner Mongolia	886.2	Henan	1,620.60
4	Jiangsu	3,175.50	Henan	573.2	Inner Mongolia	1,169.00
5	Sichuan	3,140.00	Ningxia	539.5	Anhui	575.5
6	Anhui	3,023.30	Anhui	494.2	Shandong	512.8

Source: 2008 China agriculture statistical report

1.2 Favorable Ecological Environment

The winter is bitterly cold in Heilongjiang Province, and summer temperatures are low too. For that reason, the growing season for agricultural crops is short, but there are few kinds of disease and pest damage. The amounts of pesticides and chemical fertilizer used per unit area were just 53.6 and 52.2 % of the national averages.

The percentage of the forested area of Heilongjiang Province is more than twice the average for the whole country. The history of agriculture is comparatively short, but the greater part of agricultural land is fertile black earth, and the organic content in the soil is great. Moreover, as the population density is only 61.8 % of the national average figure, industrialization in rural areas has not progressed and environmental pollution is low. Furthermore, Heilongjiang Province is located in northern China which has abundant water resources, and the environment for agricultural production is suited to the production of good and green food. Yet further, the growing of genetically-modified agricultural crops has been strictly prohibited (Jiao 2008, pp. 71–80).

1.3 Few Hours of Sunlight

Heilongjiang Province, which is situated between 43° 25′ N and 53° 33′ N, has a monsoon continental climate, its daylight hours are few, and its cold-region climate is characteristic. Consequently, it is suitable for the production of crops with a short growth period, and the harvesting period for crops is concentrated into once a year. In addition, with the change in the amount of precipitation being large, depending on the season, as most of the agricultural land lies on a plain, disasters such as drought and flooding frequently occur.

1.4 Large Disparities Among Regions

The area of cultivated land per rural household in Heilongjiang Province, at 2.4 km², is equivalent to four times the national average figure. The disparities in the scale of production among areas within the province, however, are large. For example, the population density of the western Songnen Plain is relatively high within the province, and its scale of production is small. On the other hand, the population density of the eastern Sanjiang Plain is relatively small, and its scale of production is large. In particular state-owned farms account for approximately one quarter of the total cultivated land in the province, and, distributed in the east and north, the scale of production per household is equivalent to 7–8 times that for other areas. In contrast to many general farmers having one-man operations, organizational management has become the case for farms. Because a shift in farmers' land contract management rights is being encouraged, the disparities in the scale of production are large even for farmers in the same area.

1.5 The Limits to the Transportation Capacity for Agricultural Products

Heilongjiang Province is situated in the northernmost part of China, the transportation distance to the consumer markets for agricultural products is great, and the transportation costs are high. Railways have mainly been used for the transportation of staple commodities, including food, but as the volume transported is large for Heilongjiang Province for coal, lumber, and oil, etc., there is a limit to the rail transportation capacity. Consequently most food harvested intensively gets constrained by the rail transportation capacity, and the regularity and stability of the transportation of agricultural products out of the province have not been secured. Currently, plans for rail transportation, road transportation, integrated marine and river transportation, and sea–land intermodal transportation are being formulated.

2 The Trends in Agricultural Production Development

2.1 The Tendency Towards Centralization of Crop Production

In recent years, the area of land in Heilongjiang Province used for food cultivation has increased rapidly, reaching 12.95 million hectares in 2009, and constitutes 93.8 % of the total area for crops. The commodities which have increased are mainly maize, soybeans and rice. These three together account for 93.2 % of the area for food cultivation, and have increased 42 % over the last 30 years (Fig. 12.1).

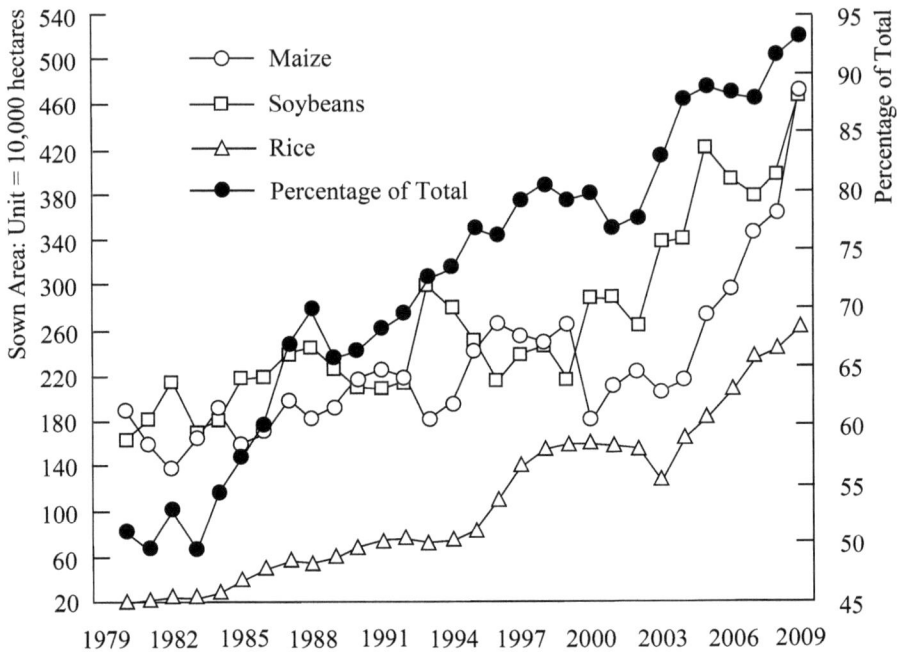

Fig. 12.1 Area of major crops and percentage area used for food

For these three crops a tendency towards centralization regionally can be seen: maize is mainly unevenly distributed in the southwestern region where temperatures are relatively high, soybeans are mainly concentrated in the northern region where temperatures are relatively low, and rice is mainly distributed in the southern and eastern regions where water resources are relatively abundant.

2.2 The Emphasis on Quality

In accordance with the opening-up of markets for selling food, and depending on the use for the food, the sales routes, and the quality, a great gap in the sales price of food has also occurred. Based on the price-setting according to the quality of products, the sales price for distinctive varieties, including select varieties of rice, high starch-content maize, and soybeans for food use, have been set relatively high, production has shifted to a stage concentrating on specialized products, and agricultural production is transforming from a "quantitative increase" to a "pursuit of high returns".

2.3 The Integration of Operations

The Chinese government buys up food at the minimum guaranteed price, but farmers can sell their produce on a voluntary basis. Besides state-owned cereal warehouses, there are also such sales routes as individual sales and processing by firms. In particular, the food production method of the "integration of production, processing, and sales", which is centered on firms, is on an increasing trend. For farmers, the sales price to food-processing firms is higher than the government purchase price. Moreover, a shift can be seen from individual management by farmers to the organization of agricultural production and to the development of technological services.

3 The Current Status of Agricultural Industrialization

3.1 Low Processing Capacity

Currently, sales to state-owned cereal warehouses have become the principal route for farmers' food sales. For this reason, most food that is bought up remains unprocessed, and in many cases is transported out of the province. On the other hand, there are many food processing firms in Heilongjiang Province, but there are few large-scale firms. For example, in 2008 Heilongjiang Province rice processing firms exceeded 2,600 in number, but firms which have an annual processing capacity of more than 25,000 tonnes are just 20 % of the total, and the average utilization rate of processing facilities is approximately 49 %. In other words, the supply-chain structure for food processing is short, and processing capacity is relatively low (Heilongjiang Provincial Development and Reform Commission 2008; Heilongjiang Provincial Grain Bureau 2009).

3.2 Rapidly Increasing Large-Scale Firms

Large- and medium-scale food-processing firms are on an increasing trend, and the processing forms are shifting from simple processing to high-value-added processing and by-product processing. Such firms as "Jiusan Grain & Oil Industry Group Co., Ltd.", "Beidahuang Rice Industry Group Co., Ltd.", and "Zhongliang Biochemical Energy Resources (Zhaodong) Co., Ltd." are growing into relatively large-scale processing firms in national terms. Twenty companies in Heilongjiang Province have already become leading agricultural industrialized food-processing firms at the national level. In particular, firms have also increased which have received investment by a multitude of routes from outside the province. In the case of rice-processing fierce interregional competition can also be seen (Zhang et al. 2009, pp. 10–15).

3.3 Improvement of Production Mechanisms Is Necessary

The "integration of production, processing, and sales" method by firms, processing bases, and farmers, continues to be the basic model for agricultural industrialization. Its main objective, by means of an "order-made agriculture", is to resolve the issue of a "muddled complexity of many varieties" in agricultural production. Within this method, however, there remain problems such as "the relevance of agricultural production materials (agricultural tools and fertilizer, etc.) is low", "effective technical support for production is small", and "profit-sharing mechanisms are unclear". Other than this, the Chinese government's minimum guaranteed price policy is also having a great impact on processing firms buying up foodstuffs.

3.4 Insufficient Brand Effectiveness

Processing firms have expectations for brand effectiveness for agricultural products, but the systems requiring the production and quality of green food products and organic food products are complex. For the country as a whole, Heilongjiang is the province with the greatest area for production of green food products and the largest number of brands. However, the number of brands may be large, but the issue of well-known brands being few and brand strength being low still remains. The minuteness of firms' scale of production and the inadequacy of their production mechanisms is having a great impact on brand effectiveness.

4 The Challenges for Agricultural Industrialization

4.1 Selection of Quality Varieties

Seen from the perspective of a favorable ecological environment, Heilongjiang Province is best suited for agricultural produce processing industries involving food. At the same time, when selecting agricultural crop varieties, there is an advantage for varieties which are adapted to cold regions. By consolidating with quality agricultural crops it is possible to expand the scale of firms. As shown in Table 12.2, agriculture in Heilongjiang Province, besides the principal crops such as rice, maize, and soybeans, is also advantageous for autumn vegetables including potatoes,

Table 12.2 Other preponderant crops in Heilongjiang province

Crop	Area	Percentage of national total
Miscellaneous beans	28.79	9.63
of which red beans	6.59	32.51
Potatoes	24.49	5.25
Sunflowers	10.73	11.13
Beet	9.04	36.69
Barley	7.64	9.63
Flax	3.64	64.2
Watermelons	5.99	3.46
Muskmelons	3.61	9.98

Source: 2008 China agriculture statistical report
Units: 10,000 ha; %

sunflowers, beet, and flax. Among autumn vegetables also, great potential lies in the processing sector for Chinese cabbage, cabbage, carrots, radishes, green onions, garlic, and eggplants, etc.

4.2 Sales

Heilongjiang Province has a population which is relatively low, and a large land area. The agricultural produce processing industry must put effort into sales outside the province. The securing of stable sales markets is the key to the development of the agricultural produce processing industry in the future. To that end, methods including the linking of production and sales, the combining of firms and sales networks, and cooperation between local firms and overseas firms will be imperative for the development of agricultural industrialization (Shi 2009).

4.3 The Differentiation of Manufactured Goods

Through emphasizing the special characteristics of manufactured goods, further enhancing the precision of processing, and aiming at consumer markets at home and overseas, it is possible to further promote the industrialization of agriculture (Table 12.3). Heilongjiang Province's soybean production volume is approximately 40 % of that for China in its entirety, and its commoditization rate has exceeded 80 %. Imported soybeans used for edible oil, however, have grown to 2.3 times the total soybean production volume for China, and in addition are on an increasing trend. If they strive for differentiation, including that of making Heilongjiang Province's soybeans for direct food use, the likelihood for developing a highly profitable

Table 12.3 Main crop characteristics and advantages

Crop	Total production	National	Percentage of national total	Characteristics of produce
Soybeans	620.5	1,554.20	39.92	Soybeans for food use
Maize	1,822.00	16,591.40	10.98	Maize for food use
Rice	1,518.00	19,189.60	7.91	Round-grained glutinous rice

Source: 2008 China agriculture statistical report
Units: 10,000 tonnes; %

industry will be high. If they exploit the advantages of yellow soybeans, the ecological environment, and the non-use of genetically-modified crops, there will be the possibility to develop a large market at home and overseas (Jiao 2009). The proportion which the rice production volume of Heilongjiang Province accounts for in the national total is not high, but the commoditization rate has risen to more than 60 % of the province's production volume, and it is the largest non-glutinous rice production area in China and the world. Wuchang City in Heilongjiang Province, exploiting the characteristics of select varieties, is producing the "*Daohuaxiang* [Rice Blossom]" brand of rice. Its price is approximately 70 % more expensive than ordinary rice, and has become a model example which has achieved an increase in the incomes of farmers and a good return for firms (Chen 2007).

4.4 Innovation of Production Systems

In order to realize large-scale agricultural industrialization, firms have become the central focus, and production models should be created which integrate production, processing and sales. In order to enlarge firms, based on the current situation of agricultural produce via farmers' independent management and sales in Heilongjiang Province and on the distinguishing characteristics of its cold-region climate, they must broaden the "regional clustering production model". This model is one which selects climatically-suitable areas, and the supporting sectors of processing firms, food warehouses, and agricultural technology work together. That is to say, food warehouses make contracts with the administrative authorities of townships and villages, and take charge of buying up agricultural produce, and the administrative authorities of townships supervise the production of the farmers, the supporting sectors of agricultural technology undertake technological support, and firms take charge of processing and sales. Through this, they will solve the problem of all kinds of varieties, realize the standardization of production technology, and it will be possible to encourage the participation of farmers in the expansion of the scale of production, by way of rational profit-sharing mechanisms (Jiao et al. 2007).

5 Conclusion

What can be raised as the principal characteristics for the agriculture of Heilongjiang Province is that the rate of commoditization of food is high and the volume of production is on an increasing trend. Agricultural industrialization is still at the initial stage, but its pace of development is fast, and the means of investment shows a diversified appearance. For the food processing industry primarily, as shown by the slogan "play the green card and travel a unique road", if they are able to enhance the precision of processing, aiming at consumer markets at home and overseas, there will be great potential for the agricultural industrialization of Heilongjiang Province.

References

Chen B (2007) "Live Rice" enters the home, and "Meiyu" husked rice sells at an astronomical price. Heilongjiang Rural News Daily, 30 Nov 2007, First edition [in Chinese]
Heilongjiang Provincial Development and Reform Commission (2008) Plan for the building of grain production capacity of fifty million tonnes in Heilongjiang Province [in Chinese]
Heilongjiang Provincial Grain Bureau (2009) Development plan for the construction of Heilongjiang Province Rice Processing Park [in Chinese]
Jiao J (ed) (2008) Rural economic development and increasing the incomes of farmers. China Agriculture Press, Beijing [in Chinese]
Jiao J (2009) China should emphasize the development of soybeans for food use. Soybean Sci Technol 2009(3):3–4 [in Chinese]
Jiao J, Xu X, Nakamoto K (2007) A new model for consolidated production in rice areas. North Rice 2007(4):56–58 [in Chinese]
Shi Z (2009) Longjiang rice dispute played out. Heilongjiang Daily, 27 Oct 2009 [in Chinese]
Zhang F, Liu Y, Fu J (2009) Research on the development strategy for the grain and oil processing industry of Heilongjiang Province. Heilongjiang Grain 2009(5):10–15 [in Chinese]

Chapter 13
Agricultural Production and Related Business by Public Firms: A Case Study on Xinhua Farm, Heilongjiang

Hironori Yagi and Yonghao Zhu

Abstract We evaluated the current status of rice production in Heilongjiang Province based on our survey interviews of Xinhua farmers. First, there is the problem of the gap between farmers on state and conventional farms. The scale of conventional farming is small, and suffers from so-called structural problems. Social unrest will increase with further widening of the gap between the rich and poor. If a subset of farmers moves to large-scale operations with machineries, then securing employment opportunities for the remaining farmers will become an issue. Second, farmers will have difficulties in achieving a further increase in the volume of production by yield increase and farmland reclamation. Third, regarding rice-planting machines and driers, while current performance is not high, their gradual but growing use has a high potential to increase operational efficiency. Fourth, because farmers use few organic fertilizers such as compost, and depend on chemical fertilizers, profitability will worsen as prices for these resources increase. Fifth, labor costs will trend upward with the rise in GDP. Sixth is the problem of water resources and securing prime agricultural land. The demand for water from cities and factories is high, and it is possible that water resources will become tight in the future.

Keywords Rice production • State farm • Heilongjiang Province

---

H. Yagi (✉)
Department of Agricultural and Resource Economics, Graduate School of Agricultural and Life Sciences, The University of Tokyo, Tokyo, Japan
e-mail: ayouken@mail.ecc.u-tokyo.ac.jp

Y. Zhu
Faculty of Economics and Business Administration, Fukushima University, Fukushima City, Japan

1 Purpose of This Chapter

Heilongjiang Province, northeastern China, is one of the country's leading grain-producing regions, particularly of Japonica rice. However, its rice-production history is relatively recent.

Following the reforms that came with China's Open Door Policy, Heilongjiang's total arable land increased from 8.46 million ha in 1978 to 9.62 million ha in 2000. Over the same period, the percentage area of irrigated land expanded from 3.2 to 17.1 % (Wang et al. 2003). The province also experienced a remarkable increase in land productivity. The rice yield, which was 2–3 tons/ha or more from the 1950s through the 1970s, rapidly increased in the 1980s, reaching approximately 7 tons/ha by the end of the 1990s.[1]

We can classify farm management entities in China into two broad types based on land ownership and managerial decision-making authority. One type is private management by private individuals, while the other is management by the state (nongken: "state-owned land-reclamation enterprises" or "state farms"). The former, through the implementation of the production contracting system, was implemented in 1984, and allocated 30-year contract (land-use) rights from the People's Communes to individual farmers, while maintaining the collective ownership of the land.[2] The latter, on the other hand, implemented a production contracting system on state-owned land. Here farmers were employed in state farm enterprises as "workers" under the specific guidance and supervision of farm executives who managed production (note that most workers were unsalaried and paid rent). Heilongjiang's state farms were established in 1949, and centered on the return of the People's Liberation Army to collective farming (rongjun nongchang) and the liberation of Kuomintang Army prisoners (jiefang tuan nongchang). By the end of 1955, about half of Heilongjiang's present farms were established, accounting for 225,000 ha (3.3 %) of the province's arable land area and 0.9 % of the provincial farm population. Subsequently, the return of military personnel and youth to farming further expanded the state farms' share of agriculture in Heilongjiang, reaching 22.5 % of the provincial arable land area and 15.7 % of total agricultural production by 1995.[3]

Nine branch offices under the Heilongjiang State Farm Division guide individual farmers and ranchers. Each farm not only produces farm products but also manages factories, hospitals, and schools. Indeed, state farms form a sort of "town" on the reclaimed land. While the control of agricultural production is left to individual farm management, the state-owned sector handles purchasing, processing, sales, and exports.[4]

---

[1] Park and Sakashita (1998), p. 92.
[2] See Park and Sakashita (1996), Zhou and Abe (2000), and Sun (2003) for the issues for private farms in Northeast China.
[3] Park and Sakashita (1998), pp. 92–99. Also, see Dong (1998) for details of state farms.
[4] Park and Sakashita (1998), pp. 93–94.

By the end of 2007, the population of the state farm division was 1.65 million, agricultural acreage increased to 2.39 million ha (including 2.162 million ha in food-crop acreage), and agricultural production (including potatoes) reached 12.46 million tons.[5]

In early September 2008, we conducted an interview survey at a state farm enterprise in Xinhua.[6] In particular, we visited the Heilongjiang Beizhu Rice Co., Ltd. (hereinafter Beizhu Rice), which coordinates Xinhua Farm production regiments and sells rice from the farms. In addition to conducting interviews on the general conditions of production and sales, we visited State Farm Harbin's branch office and the Northeast Asia Research Station to collect information on state farm enterprises in Heilongjiang Province. In this chapter, based on the results of our interview survey for Xinhua Farm, we analyze the state of agricultural production and related businesses and present a vision for the future of rice production.

2 Survey Results for Xinhua Farm

2.1 Position of State Farm Enterprises in Heilongjiang Province

According to the interviews conducted with the relevant authorities, Heilongjiang's state farm enterprises account for approximately a quarter of the food crop area under production in the province, approximately one third of the agricultural production, and half of the commercial agricultural products. In particular, state farms produce 60 % of the province's commercial rice.

On average, farm households on state farms are 10 ha in size, although some are over 300 ha. These farms enhance efficiency by using fertilizing machines and aerial pesticide control, and increase the yield to 30–50 % per unit area. The state owns the state-farm land, which accounts for most of the land reclaimed over the last half century. In recent years, however, the government had to regulate land development due to land erosion and water shortages. Barriers to entry into the state farm also exist. Before one can become a state farmer, one must become an employee of the state farm. Moreover, when a farmer purchases heavy machinery, the government provides financial assistance of 20–30 %. Although some farmers work under contract using their own machinery, at present, few farmers own the machinery.

In contrast, Heilongjiang's conventional farm households, outside the state farms, have an average area of 2 ha and number approximately four million. Many conventional farms have 30-year contracts that give them agricultural land-use rights, and they are free to decide on which seeds, pesticides, and machinery to use

---

[5] See pp. 43–59 of Heilongjiang Province State Farm Directorate Bureau of Statistics (2008).
[6] For details of the survey interview, see Zhu (2009). Regarding the situation of Xinhua Farm, the following paper had already presented analysis results: Park et al. (2001), pp. 85–98.

Fig. 13.1 Xinhua Town center

among other farm-related issues. Manual work predominates; therefore, pesticides are not used. As a result, input costs are lower on these conventional farms.

2.2　Realities of Production on Xinhua Farm

Xinhua is located in the vast Sanjiang Plain, at the confluence of the Amur, Songhua, and Ussuri Rivers. The farms are named after their location in Xinhua Town, Dongshan District, Hegang City, Heilongjiang Province (Fig. 13.1).

At the end of 2007, Xinhua Farm had a total population of 23,266 and included 8,615 farm households. The employees numbered 12,717 (of which 9,049 were farmers). The total farm area was 55,873 ha and the agricultural land area was 29,307 ha (11,333 ha of which were rice paddies). The main crops were rice, soybeans, corn, wheat, and barley. Cereal production reached 152,469 tons in 2007, of which rice and soybeans accounted for 96,267 and 8,225 tons, respectively.[7]

---

[7] Heilongjiang Province State Farm Directorate Bureau of Statistics, op. cit. (Footnote 5), pp. 341, 375–379.

About half of the Xinhua farm households were 100 mu (1 mu = 6.7 a) and under. Production was divided into regiments, with each regiment of approximately 1,000 ha in size and comprising 100 households. According to our interviews, large-scale farms were 16–17 ha or more.

The cost of renting land on a Xinhua farm is 3,300 yuan/ha. There is also an additional water fee of 300 yuan/ha, according to interviews with regimental leaders. The rice yield is fairly large at 7–7.5 tons/ha. The yield of soybeans is 3 tons/ha. The average household on the farm earns a salary of 3,000–4,000 yuan/ha, which assures a good standard of living. A typical farm household on the state farm does not receive a salary, although a farm's regimental leader (lianduizhang) receives a salary of 20,000 yuan/year. Further, it is compulsory to participate in crop mutual aid as insurance in case of failed crops or other calamities.

Rice paddies as large as 1 ha exist, although such large plots are discouraged and it is thought that 0.5 ha is a reasonable size. Farm roads are gradually becoming two-laned and paved, as can be seen in Fig. 13.2, and by 2009 all regiments were to have upgraded their roads to this standard.

It was not general practice to use compost on Xinhua Farm, although it was used on part of the farm on a trial basis. Rice seed was centrally managed by the farms.

To ensure a long growing season, farmers are directed to plant their fields by 20 May. Rice-planting machines planted 95–100 % of the paddies and mechanization

Fig. 13.2 Paved farm road

Fig. 13.3 Rice harvester (simplified type, six rows) in storage shed

has continued. Almost all farm households possess a rice-planting machine and there are common-use sheds where a simplified type of rice-planting machine (six rows) can be stored, as seen in Fig. 13.3. These rice-planting machines cost approximately 12,000 yuan each and have a working capacity of 2 ha/day. The rice-planting season is the busiest, and the upper limit for one person working alone is approximately 5 ha of rice acreage. After the planting season, most farmers work as day laborers in other sectors and earn about 100 yuan/day as temporary income.

All crop protection on the farm is conducted by aerial spraying, at a cost of 150 yuan/ha, and is usually done twice a year. Within the cost of crop protection, 70 yuan goes to airplane rental and 80 yuan covers pesticide costs.

Almost all harvesting is done by machine (95–100 %), but this does not mean all farmers own a combine harvester. Those who own combine harvesters also work under contract. Contract work fees at the time of interview was about 600–700 yuan/ha, increasing to about 1,000 yuan/ha for rice fields in which rice stalks had fallen over, thus causing more work. It is possible to harvest 10 ha/day, with a working capacity of 10 h/day. One of the regiments (liandui) has 20 2-m-width-class harvesters and 7 3-m-width-class harvesters (which cost about 230,000 yuan) (Fig. 13.4). They also have a head-feeding combine. In addition, service businesses exist that perform combine harvesting under contract. It is said that if farmers save for 2–3 years, they can purchase a machine and redeem their investment within 4 years.

Fig. 13.4 Combine harvester in storage shed

2.3 Processing and Shipping: Results of Our Interview Survey on Beizhu Rice

At the time of our study, Heilongjiang's Beizhu Rice, which solely handles the rice production of Xinhua Farm, was a Japanese–Chinese joint venture that dried, stored, milled, and sold rice (Fig. 13.5). Beizhu Rice has a capital of 6.05 million yuan, with total assets of 230 million yuan. Major stakeholders are: Sojitz, a Japanese trading company with a 25 % stake; Tsuruoka Rice Processing Co., Ltd., with a 65.6 % stake; and Heilongjiang State Farm Division, which has a 9.4 % stake. Beizhu Rice at the time had about 40 full-time, permanent employees, with 36 people outside of the management section in a double-shift system. In addition to a group chief, each group had 18 members: 1 person as a leader, 12 persons in product processing, 1 person in charge of inspection, 1 person in material procurement, 1 person in maintenance, 1 person in product management, 1 person in government relations, 2 persons in finance, and a number in sales. In addition, there were part-time workers who handled simple tasks, and the facilities were often run at night, because electricity rates were cheaper during these off-peak hours.

All 380 contract farmers are on Xinhua Farm and have a total area of 4,200 ha under contract. These contract farmers ship all rice meant for sale to Beizhu Rice. Being particularly thorough on cultivation methods, farmers are brought together for a meeting before planting in order to be instructed on how to use fertilizer and pesticides. It is said that non-Xinhua farmers are not contracted because they do not follow the prescribed practices (Fig. 13.6).

Fig. 13.5 External view of Beizhu Rice Processing Co., Ltd

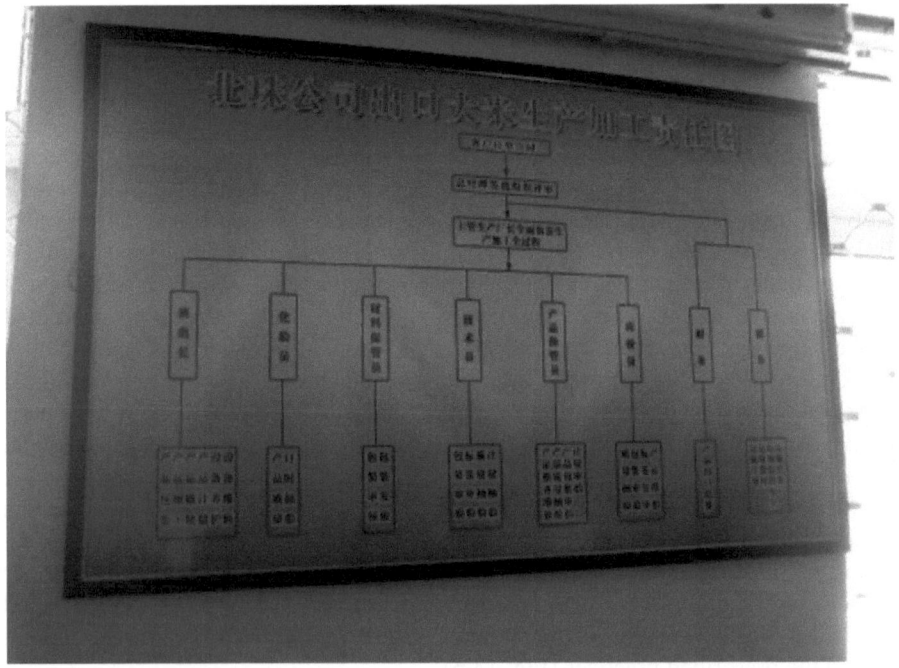

Fig. 13.6 Organizational chart of Beizhu Rice Processing Co., Ltd

The farmers can only acquire designated varieties of rice, grade 3 and above. Rice varieties consisted of Soraiku 131 (which accounts for one third of all varieties), Soraiku 163, and New Koshihikari. Farmers pay an extra 0.05 yuan/500 g for Soraiku 131, which is a top-rated variety. However, farmers sometimes do not want to plant it as it falls down easily and is susceptible to damage from the weather. The usual price to buy grade 3 rice from farmers is 1.8 yuan/kg, but a premium of 0.02 yuan is added for grade 2 rice.

Harvested rice paddy is delivered in October and left at the entrance of Beizhu Rice to dry in the sun. It is then temporarily stored in the silo shown in Fig. 13.7 (storage capacity of 200 tons). Next, it is placed in the back of the mill and dried in a drier. The mill has two Kaneko driers with a capacity of 10 tons each.

Of the dried brown rice, the portion under a sales contract is immediately milled. There are two milling machines: one 14.4 kl (made by Satake for 1.2 million yuan, installed in 1998) and one 16.2 kl (made by the Anzai Manufacturing Co., Ltd. for 600,000 yuan, installed in 2001). Beizhu Rice mills 100 tons of rice daily, and ships about 30,000 tons every year. There are eight new silos for storing the dried brown rice (Fig. 13.8), but because they only have a total storage capacity of 2,000 tons, Beizhu Rice also stores brown rice outside in the winter.

The processed rice is exported abroad via the China National Cereals, Oils and Foodstuffs Corporation (COFCO). This is China's largest food company, and it holds export rights. Until 2007, foreign markets accounted for 60 % of rice exports, beginning with Japan at 5,000 tons, then Hong Kong at 3,000 tons, and the rest

Fig. 13.7 Temporary storage facility for unprocessed rice

Fig. 13.8 Storage facilities for brown rice

going to Southeast Asia, Britain, and other countries. The remaining 40 % of production is for mainland China. However, in September 2008, because the government enacted export regulations, COFCO actively searched for domestic markets to which they could ship rice, and consequently profits were cut in half. The company hoped to return profits to prior levels; however, in addition to the government not providing special compensation, domestic prices did not increase much, and stand at about 3,000 yuan/ton.

3 Current Status of and Future Challenges for Rice Production in Heilongjiang Province

We evaluated the current status of rice production in Heilongjiang Province based on our survey interviews of Xinhua farmers.

First, in recent years we have seen rapid growth or improvements in labor costs, land rent, machinery investment, roads, and rice quality. Among these, we expect labor costs to adjust with economic growth and to approach those of advanced nations.

Regarding machinery, while almost no farmers owned combine harvesters in the 1990s,[8] our 2008 investigation revealed that the use of combine harvesters has become fairly common and that highly efficient machines have also been introduced. However, farmers still have simple rice-planting machines, and maintenance standards for drying and storage facilities are insufficient given the amount of production. Finally, although farmers lack sufficient drying and storage facilities and almost none use organic fertilizer, these issues are improving, at least in regards to rice quality.

Second, some regions lagged in aspects such as yield per unit area and agricultural land development. The rice yield had already risen to the 7 ton/ha level 10 years ago, and it is difficult to imagine that this yield will increase sharply in the future. Further, it is becoming difficult to find land suitable for new agricultural development; where such land is available, it is hard to find resources to develop it.

Third, small-scale conventional farmers remain stalled at a low standard of living. Unlike farmers on state farms, conventional farmers do not own many machines, and because the area of arable land they manage is small-scale, their labor efficiency and income levels are low.

Fourth, the factors that could inhibit the maintenance and expansion of rice production in the future include water resource shortages and price increases for inputs such as oil and other resources.

In light of these points, and from the standpoint of rice production in Heilongjiang Province, we may make the following observations regarding the future prospects for food security in the Northeast Asian region.

First, there is the problem of the gap between farmers on state and conventional farms. The scale of conventional farming is small, and, like Japanese farms, conventional farms suffer from so-called structural problems. Also, some are concerned that social unrest will increase with further widening of the gap between the rich and poor. Thus, if a subset of farmers moves to large-scale operations and increased mechanization, then securing employment opportunities for the remaining farmers will become a major issue.

Second, farmers will have difficulties in achieving a further increase in the volume of production. As noted earlier, because we cannot foresee an increase in yield and crop acreage, expanding the value of agricultural production through an increase in the volume of production might be difficult in the future.

Third, regarding rice-planting machines and driers, while the performance and ownership of this machinery is not high, their gradual but growing use has a high potential to increase operational efficiency.

Fourth, because farmers use few organic fertilizers such as compost, and depend on chemical fertilizers, profitability will worsen as prices for these resources increase. Consequently, further value added will require farmers to reduce the use of chemical fertilizers and pesticides.

---

[8] Park et al. (2001), pp. 93–95.

Fifth, labor costs will trend upward with the rise in GDP. To counter this, there are plans to increase the labor productivity of farmers on state farms by enlarging rice paddy partitions, introducing large machinery, and scaling up the acreage of arable land managed.

Sixth is the problem of water resources and securing prime agricultural land. The demand for water from cities and factories is high, and it is possible that water resources will become tight in the future.

Further, from the viewpoint of food security in Northeast Asia, the significance of Japan's rice paddy agriculture is high with regard to water resources and land suitable for agriculture. When we compare the rice paddy agriculture of Heilongjiang Province to that of Japan, Japan's wage level is still high, although the gap is gradually shrinking. On the other hand, Japan has been transferring rice varieties and cultivation technology to China and elsewhere, thereby helping Heilongjiang's state farms achieve very high rice yields. It can be difficult to judge overall quality, but as of September 2008, the probability that Japanese consumers would prefer the taste of rice produced in Japan was high. In the future, along with securing prime agricultural land and expanding the scale of managed arable land, Japan will need to promote agriculture that addresses the depletion of oil resources and uses lower amounts of chemical fertilizers and fossil fuels.

This chapter is the product of research done by the Economic Research Institute for Northeast Asia (ERINA) undertaken in fiscal year 2008 in "Food Security for Northeast Asia (2)"; the manuscript was reproduced in ERINA Report, Vol. 88, published in July 2009.

References

Dong Y (1998) Issues and realities of the farm head responsibility system in China's state-owned farms. J Rural Econ 1998:273–276, Special issue [in Japanese]

Heilongjiang Province State Farm Directorate Bureau of Statistics (2008) Heilongjiang Province reclamation ward statistical yearbook, 2008 edition.

Park H, Sakashita A (1996) The development of individual farm management and coordination of land use in northeast China. J Rural Econ 1996:208–213, Special issue [in Japanese]

Park H, Sakashita A (1998) Characteristics of farm management reform in northeast China. Rev Agric Econ 54:87–100 [in Japanese]

Park H, Sakashita A, Da Z, Yoshida K (2001) Rice paddy development and management on state-owned farms in Sanjiang plain, China: a case study of Xinhua Farm. Rev Agric Econ 57:85–98 [in Japanese]

Sun X (2003) Management analysis on rice crop agriculture in Heilongjiang province. Jpn J Farm Manag 41(2):134–137 [in Japanese]

Wang H, Lei G, Nakagawa M (2003) Factors and changes in the use of agricultural land in China's leading food production base: a case study of Heilongjiang province. Jpn J Farm Manag 41(2):129–133 [in Japanese]

Zhou S, Abe J (2000) Consideration of rice production in Heilongjiang, China. J Rural Econ 2000:251–255, Special issue [in Japanese]

Zhu Y (2009) Observation report on Heilongjiang Province agricultural production and farm management. ERINA Rep 85:49–54

Chapter 14
Promotion Policies for Food Industry Cluster in Korea

Byung-Oh Lee

Abstract The Korean government has been developing a National Food Cluster in Jeollabuk-do Province since 2008 to lead the development of regional small scale food industries, and to expand R&D capabilities within the food industry. In this complex, a Food R&D Research Center and other supportive facilities will reside and be expected to serve as a hub in the development of the Korean food industry clusters. To develop the food industry in the Gangwon-do Province, various differentiated clusters should be implemented with unique items and functional ingredients based on the regional characteristics. Four types of clusters were suggested, such as item based cluster, functional cluster, traditional food cluster, and local food cluster. Also, by interconnecting a vast tourism infrastructure and bringing tourists into local food industries, the food industry range might be expanded and produce a synergistic effect from mutual complement relationships between tourism and food industries. Traditional and local food should be differentiated by not only its quality and functionality, but also in design as well, and this requires advanced technical support. To achieve this, it needs to install a Food Development and Technical Support Center within the Gangwon-do Province Agricultural Research and Extension Services.

Keywords Food industry cluster • Functional cluster • Local food cluster • National food cluster • Traditional food cluster

---

B.-O. Lee (✉)
Kangwon National University, Chuncheon, Republic of Korea
e-mail: bolee@kangwon.ac.kr

© Springer Japan 2016
L. Kiminami, T. Nakamura (eds.), *Food Security and Industrial Clustering in Northeast Asia*, New Frontiers in Regional Science: Asian Perspectives 6, DOI 10.1007/978-4-431-55282-6_14

1 Introduction

The Korean agricultural industry has greatly contracted since the launching of the World Trade Organization (WTO) in 1995. In addition, the Free Trade Agreements (FTAs) with Chile, the European Union, and the United States are accelerating the liberalization of markets. The current condition of the South Korean agricultural industry, with relatively small cultivation acreage per family, is limited in its ability to expand the scale of operations in order to reduce production costs. As a result, liberalization in South Korea is closely linked to the stagnation of agriculture income, and the current situation reflects growing gaps between farming incomes and urban working incomes.

To overcome these difficulties, the Korean government reorganized the Ministry of Agriculture into the Ministry for Food, Agriculture, Forestry, and Fisheries in 2008 (later renamed Ministry of Agriculture, Food, and Rural Affairs, MAFRA in 2013). Much effort is being made to promote and enhance competitiveness in the South Korean food industry. Currently, the Food Industry Policy Unit under the Ministry of Agriculture, Food, and Rural Affairs oversees various aspects, including food industry policy, food industry promotion, food service industry promotion, export promotion, marketing policy, and consumption policy. Additionally, the Food Industry Promotion Act was enacted in 2008 and renewed the Fundamental Law of Agriculture, Fisheries, Food Industry, and Rural Areas to establish a legal foundation for promoting the food industry.

The food industry contributes to increasing value on agricultural products supplied from rural areas, and also holds great importance in aspects of employment, exportation, and more. This is especially the case for traditional foods made from domestic agricultural products, which have less competition with imported food products and more continuous demand.

However, the food industry requires capital and technology, and must be equipped with safety management systems. The government's political support and R&D network of related industries are crucial for the development of the food industry. In that regard, a food industry cluster is an effective method to promote the food industry in general, and many advanced countries have already implemented this type of system. A food industry cluster is conceptualized as a group of food companies, universities, and laboratories within a specific area which establish a network, and through their interactions create a synergistic drive for technology research and development, ingredient procurement, effective human resource management, and information sharing within an organizational system.

The South Korean government was aware of the food industry cluster concept and initiated the creation of the National Food Cluster (NFC) beginning in 2008. The government has also been conducting small-scale regional food industry cluster promotion projects (the Regional Strategic Food Industry Promotion Policy) since 2005. However, the history of these projects is short and their results are minimal, so it has yet to be evaluated. The purpose of this research is to take a general overview of the structures and features of the South Korean food industry and introduce food industry cluster promotion policies.

2 Structure and Features of the South Korean Food Industry

2.1 The South Korean Food Industry Structure

The South Korean food industry has been in continuous development since the 1980s when industrialization advanced alongside economic growth and increases in income. Table 14.1 below shows the food industry revenue in 2012 was 152 trillion KRW, which is a 91.9 % increase from 2002. The industry employs 1.93 million workers, which is an 11 % increase from 2002. Production in agriculture and fisheries in 2012 accounted for 51 trillion KRW, showing a 36.4 % increase from 2002.

Table 14.2 below shows the number of food manufacturing firms in Korea (those with more than 10 employees) for 2012 was 4,423. Among these firms, operations with less than 50 employees make up 80.5 % but are only responsible for 27.7 % of the total revenue. Also, small-size sole proprietorship companies with less than one billion KRW in revenue per annum make up just 15.9 % of the revenue share from the entire food manufacturing industry.

On the other hand, large corporations with more than 300 employees make up only 1 % of the market share, but they take in 15.5 % of the total revenue. The

Table 14.1 Revenues and employment progression of Korean food industry (units: billions of KRW, thousand people)

Category	2002	2006	2008	2012
Revenues	79,430	95,274	131,291	152,435
Food manufacturers	38,934	44,381	63,725	75,150
Food services	40,491	50,892	57,566	77,285
Employment	1,741	1,600	1,780	1,932
Food manufacturers	155	150	171	179
Food services	1,586	1,450	1,609	1,753

Sources: Statistics Office, mining and manufacturer industry research and whole-sale and retail research, Annual Reports
Note: Food manufacturer refers to a company with more than 10 employees

Table 14.2 Food manufacturer status in Korea (2012)

Category	No of firms	No of employees	Revenue (one billion)	Average revenue (one billion)
Small firms (less than 50 people)	3,561 (80.5)	74,213 (41.5)	20,854 (27.7)	5.9
Mid size firms (50–300 people)	819 (18.5)	83,781 (46.8)	42,698 (56.8)	52.1
Large firms (more than 300 people)	43 (1.0)	20,845 (11.7)	11,598 (15.5)	269.7
Total	4,423 (100.0)	178,839 (100.0)	75,150 (100.0)	17.0

Source: Statistics Office (2012) Mining and Manufacturer statistics

average revenue of small firms is 5.9 billion KRW, 52.1 billion KRW for mid-sized firms, and 269.7 billion KRW for large firms.

There is a concentrated focus on major food products by large corporations. For 2012, the average revenue of the food manufacturing industry was 17.0 billion KRW, which was worth 72 % of the total manufacturing industry's average revenue of 23.5 billion KRW.

Furthermore, of the 625,000 restaurants doing business in 2012, establishments with less than five employees make up 88.6 % of the market, and those with revenues less than 100 million KRW represent 63.8 %. These companies are mostly individually owned small businesses. Revenue per restaurant averaged at 124 million KRW, which is about one-fourth the Japanese average revenue.

The infrastructure and institutional basis, which is the foundation of the food industry's development, is also very vulnerable. A lack of R&D investment resulted in low technical standards in the food industry. There are general inadequacies found in the food statistics and information infrastructure, and human resource development systems, among other things. According to 2012 standards, 18 companies have an annual revenue of over one trillion KRW; companies such as CJ (4.7 trillion), Lotte Chilsung Beverage (2.0 trillion), Nongshim (2.0 trillion), Hite-Jinro (1.7 trillion), and Ottogi (1.7 trillion).

Export revenues from South Korean agricultural and fishery products for 2012 were 8.1 billion USD, and imported merchandise revenues 33.4 billion USD, a ratio of 1:4 of exports to imports. Examples of major export products include tobacco (7.6 %, ranking 1st), tuna (7.5 %, ranking 2nd), coffee preparation products (3.7 %, ranking 3rd), sugar (3.3 %, ranking 4th), processed seaweed (2.9 %, ranking 5th), ginseng (1.9 %, ranking 8th), and kimchi (Korean pickled vegetables) (1.3 %, raking 13th). Products are exported to regions such as Japan (29.8 %), China (16.0 %), ASEAN (14.9 %), the United States (8.3 %), and the EU (5.2 %), among others. In general, the range of export-target countries and export products are very narrow.

2.2 Features of the Food Industry

First, the average size of food manufacturers and restaurants is small, and many are individually owned, which causes vulnerability in competition. However, a few large corporations have a position of market dominance. In particular, regional traditional food corporations lack competitive technology and capital, which leads to inferior product development capabilities (Lim 2012, pp. 35–36).

Second, large food manufacturers and restaurants purchase ingredients from foreign countries because domestic agricultural ingredients are expensive. Such a system results in disconnectedness and a lack of food self-sufficiency amongst the food and agricultural industries (Choi 2013, pp. 149–151). Korea's food self-sufficiency rate was 23.6 in 2012, which is one of the lowest rates among OECD members. Additionally, there is an increase in international agricultural prices and more risk factors concerning importing foreign food ingredients, but public and private response to this is insufficient.

Third, since the consumption goods in the domestic market are limited, only a few production items are highly focused upon regionally, which leads to a structure of excessive competition. Such a phenomenon results from a lack of effort in new product development in the food industry.

Fourth, recent safety issues in the domestic market caused a high sensitivity towards food safety among consumers. However, most small-scale food manufacturers are not equipped with HACCP and other safety facilities due to a lack of capital. To acquire consumer trust in the food industry, safety management must be improved, but because this will incur an increase in facility expenses in most small food firms, it is difficult to implement.

Fifth, the government's drive towards risk analysis and other advanced safety management systems is not sustained, risk communication between consumers and the government and food corporations is not smooth, and the transparency of food product distribution is highly vulnerable. These factors result in a low level of trust in food products by consumers in general. This is a large obstacle to the development of the food industry.

Sixth, the lack of both a logistics system and awareness of consumers to buy produce locally indicates that the local food system, like that of Japan or Western countries, is still unsettled. Also, the dietary and consumer education systems are also insufficient, resulting in a minimal national understanding of the food industry.

Finally, South Korea's globalized food structure is inadequate as a result of its lack of exports of food and ingredient goods. Even though products such as kimchi and other fermented food products have been recently internationally recognized as healthy foods and their effects proven, these products are not being developed into marketable forms to be bought by foreign consumers.

3 Food Industry Cluster Promotion Policies

3.1 National Food Cluster

The Korean government is propelling the National Food Cluster project to resolve the problems of small-scale operations and lack of competitive edge for technological functions of the food industry, and to establish a higher value in the food industry with a stable supply of agriculture and fisheries. In other words, the government is promoting the food industry's R&D investment and taking the lead in the future development of the food industry.

The National Food Cluster was established in December 2008 in Iksan City, Jeollabuk-do Province. The government is planning to implement a 553.5 billion KRW budget to cover the period from 2009 to 2020 and develop a professional food industry complex with an area covering more than 3.58 million square meters. When the development project is completed, it is expected to generate 15 trillion KRW in revenue, 3 billion USD in exports, 4 trillion KRW of induced demand, and create 22,000 jobs.

This complex will be inhabited by a diversified list of international and domestic food corporations, civilian laboratories, related corporations, and supporting facilities. Currently, Chonbuk National University, Jeonju University, Wonkwang University, and Kunsan National University, among others, are located in Jeollabuk-do Province, and organizations such as the Rural Development Administration and Korean Food Research Institute will be moving near to the complex, which will greatly expand the research infrastructure.

The National Food Cluster is registered as a national industrial complex, which signifies that the government will promote the food industry as a fundamental national industry. Moreover, the residing corporations will be given incentives such as rental-fee and tax reductions, financial support for R&D investment, a convenient means for registration, and more. In addition, environmental organizations will be established, and residential space, welfare projects, and education facilities for children will attract outstanding professionals.

In the future, supportive facilities such as a National Food Cluster Support Center, and a Food R&D Research Center (the Food Functionality Evaluation Center, the Food Quality and Safety Center, the Food Packaging Center, a Pilot Plant, etc.) will also be established. The National Food Cluster Support Center will act as a comprehensive control center for the complex, and provide a variety of convenient facilities to help focus on new products for participating companies, and operate corporate support programs. Additionally, the Food Functionality Evaluation Center is responsible for researching and developing functional foods, and the Food Quality and Safety Center will deal with companies' quality of management. The Food Packaging Center will provide a customized package of support services.

Figure 14.1 below shows that the National Food Cluster will act as a hub for the national food industry where research and development will be the major focus, and strengthen cooperation and support in network logistics, industrial functions, and R&D capabilities of not only companies and laboratories in the Cluster, but also for regional food industry clusters and the Daeduk Research Complex. Within the National Food Cluster, production activities will be able to focus on higher value products (e.g., functional products, fermented food products, natural additives, etc.) and export products (e.g., ginseng, kimchi, etc.).

The Saemangeum area is a large-scale area of reclaimed land in Jeollabuk-do Province where various eco-friendly agricultural complexes, export industry complexes, and a harbor will be constructed. The National Food Cluster will be interrelated with this area to become an advanced base for food exports, targeting Northeastern Asian markets. Furthermore, a processing trade incorporating advanced technology is also planned for within the expansion.

3.2 Regional Strategic Food Industry Promotion Policy

Table 14.3 below shows that South Korea has been developing a Regional Strategic Food Industry Promotion Policy (renamed in 2011) to promote regional agriculture-related food industries since 2005. As of 2014, 77 sites have already been operating,

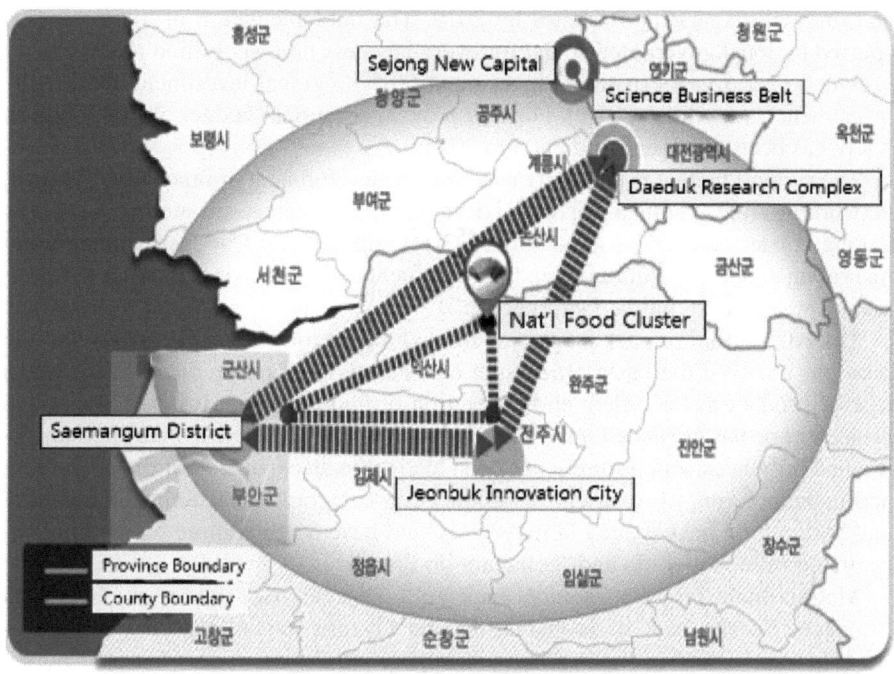

Fig. 14.1 Network of interconnected regions between national food cluster and R&D functions

Table 14.3 The trends of regional food industry cluster promotion policies in Korea

Year	2005	2008	2009	2011	2014
Title	Regional agriculture cluster		Metropolitan cluster	Regional strategic food industry promotion policy	
Sites	20	22	12	13	10
Category	Animal husbandry (18), processing (15), horticulture (14), grain (9), special crop (8), fisheries (6), others (7)				
Type	Manufacturing/logistics (35), processing (34), themed (8)				
Region	Jeollabuk-do/Jeollanam-do (13), Gyeongsangbuk-do (9), Chungcheongbuk-do/Chungcheongnam-do (7), Gyeonggi-doGangwon/Gangwon-do/Jeju (6), Busan/Gwangju (2)				
Major products	Rice, chilies, apples, grapes, tangerines, ginseng, green tea, Korean beef, pork, cheese, traditional cookies	Sweet potato, spinach, ginger, strawberries, melons, grapes, wild grapes, persimmons, figs, cornelian cherries, Korean beef, horse meet, pork, rice cakes, wine	Wheat, strawberries, persimmons, medical herbs, Korean beef, pork	Beans, wild strawberries, citrons, pork, rice processed food, processed herbal products, bibimbap, kimchi, makgeolli	Beans, tomatoes, plums, apples, turmeric, wild vegetation, pork, soybean milk, sauces

Source: MAFRA (2013a)

and this will expand to 100 sites by 2019. The implementation of this project is spurred by local governments, and for every business unit, six billion KRW will be provided over 5 years (national investment 50 %, regional investment 50 %, self-investment on facilities 20 %). For this project, a national budget of 207.4 billion KRW has been provided from 2005 to 2014.

The support budget is primarily used for the development of innovative systems, networking (e.g., human resource development systems, promoting networks among cluster entities, professional CEO recruitment, etc.), industrialization and marketing areas (e.g., mutual brand development and advertising, product development R&D support, public facilities, etc.).

Since the initial project scope was limited to certain cities and regions and the program involved the basic processing of local agricultural products, there was a lack of product development and generation of higher value. However, the recent project scope has expanded into more city and regional coalitions, metropolitan cities, and provinces, and progress has been made into an expansion of the processing and manufacturing of a variety of products. In category terms, animal husbandry and horticulture have gained larger shares, and have become more regionally concentrated in Jeollabuk-do and Jeollanam-do Provinces.

Metropolitan projects include makgeolli (Korean rice wine) in Gyeonggi Province, Korean beef in Jeollabuk-do and Gyeongsangbuk-do Provinces, eco-friendly animal husbandry in Chungcheongbuk-do Province, tangerines, a horse-meat industry, beans, and pork in Jeju Province, and wheat and kimchi from Gwangju City, among others.

The themed projects are divided into eco-friendly agriculture- and tourism-centric businesses. Eco-friendly agriculture-centric businesses include "Paldang Clean Agricultural Products" in Yongin and Yangpyeong, Gyeonggi-do Province, "Resource Recyclable Eco-friendly Agriculture" in Asan, Chungcheongnam-do Province, "Organic Rice Products" from Chungju, Jincheon and Eumsung in Chungcheongbuk-do Province, "Eco-friendly Agricultural Products" from Goesan and Danyang in Chungcheongbuk-do Province, "Eco-friendly Processed Vegetable Products" from Cheongju in Chungcheongbuk-do Province, "Revert Cycling Agriculture" from Jeongeup in Jeollabuk-do Province, and "Eco-friendly Oriental Medications" from Jinan in Jeollabuk-do Province, and more.

Tourism-centric businesses include strawberries from Nonsan, Chungcheongnam-do Province, persimmon from Wanju, Jeollabuk-do Province, "Cheongbori [green barley] Green Industries" of Jeongeup and Gimje in Jeollabuk-do Province, and wine from Yeongcheon, Gyeongsangbuk-do Province.

As the Regional Strategic Food Industry's title and promotion entity (National Food Cluster Promotion Team in the MAFRA) changed in 2011, the business has become refined, improved, and diversified into a prestigious kimchi industry, a globalized makgeolli market, organic rice food processing, a bibimbap (Korean boiled rice with assorted mixtures) industry, processed herbal products, processed bean products (e.g. soybean milk, sauces, dressings, and confectionaries), fresh ready-to-eat vegetables, fermented products (e.g. sauces and enzymes), and extracts (e.g. plums, turmeric, and wild vegetation).

Figure 14.2 below shows the operating system of the Regional Strategic Food Industry Cluster Business Units. Business Units will establish professional subsidiaries to achieve self-reliance and actual business propulsion will be carried out by the subsidiary. Of course, major issues will be decided at the Business Operation Committee. The subsidiary will receive operation and R&D consultations from the advisory committee, acquire original produce from the supplier farms, and then internally process the goods, or cooperate with other participating companies in processing, and finally undertake sales.

Although Korea's regional strategic food industry cluster has made numerous performance records for the past 10 years, issues such as poor management from insufficient professional personnel in business units, inadequate management systems such as post-evaluation systems, and lack of abilities in business model excavations, advertisements, and marketing all required improvements.

3.2.1 Pork Cluster Case

Western Chungcheongnam-do High Quality Pork Cluster in the Boryeong and Hongseong region in Chungcheongnam-do Province was established as an agricultural association in October 2009 and became an agricultural corporation in October 2010. In 2009, the Cluster began with 24 participant farms, and the number of farms expanded to 61 by 2013. In 2012, the total capital was 630 million KRW invested by 48 farms. From 2009 to 2012, 2.6 billion KRW of the national budget was put to use.

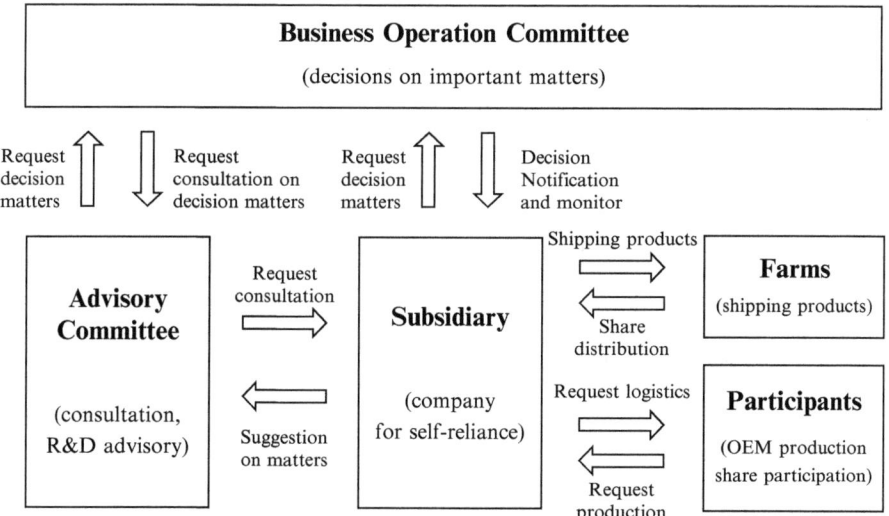

Fig. 14.2 Food industry cluster business unit operation system (Source: MAFRA (2014), p. 7)

Table 14.4 below shows the industry–academia cooperative system within the Pork Cluster. Farms are responsible for feeding and managing pork stock, firms are in charge of feed manufacture and butchery, and universities are responsible for the development of new kinds of feed and products. Manufacturing products include fresh meat, and variations of ham products, sausages, and seasoned meat. Additionally, one experiential hall and five direct outlets and restaurants (size: 300 guests per day) were also operational.

The total revenues increased 9.2 times over 2 years from 1.6 billion KRW in 2010 to 14.7 billion KRW in 2012. Profit saved up increased from 240 million KRW in 2011 to 630 million KRW in 2012. In 2012, 48 shareholders received 32 million KRW as dividends, worth 5 % of their invested capital.

Economic effects were observed in the participating farms as follows in 2012: Mutual purchase of feed and sows reduced 5 % in expenditure, feed management unification achieved a 12 % increase in productivity, and recipient price was increased by 2 % by mutual shipment, which was estimated to have about a 70-million-KRW increase in income.

For the companies, the breeding company sold 3,000 head of breeding sows (1.5 billion KRW), feed factories sold 36,000 tons of feed (18 billion KRW), and slaughterhouses had a 23,000-head slaughtering performance (5.3 billion KRW). In the universities, two patents (Omega feed, the first development of a pork traceability system in Korea) were acquired.

3.2.2 The Persimmon Cluster Case

The Persimmon Higher Value Addition Cluster was established in October 2009 as a non-profit organization in the Cheongdo, Sangju, and Mungyeong region of Gyeongsangbuk-do Province, and was switched to an agricultural corporation in

Table 14.4 Pork cluster industry-academia cooperation system

Category	Participant companies and institutions	Role
Industries	Farms: 61 farms Companies: 5 firms Nonggawon (agricultural association), Daehan feed, Dongwon farms, Samwha breeding, Hongju meat	Nonggawon: Feeding pigs Daehan feed and Dongwon farms: Manufacture of omega feed Samwha breeding: Supply breeding sows Hongju meat: Slaughter shipped pigs
Academies	Universities: Dankook University, Chungwoon University, Cheonan Yonam College	Univ.: Develop omega feed, experiment with feeding management, product development (sausage, shabu-shabu)
	Institutions: Korea Institute for Animal Products Quality Evaluation, Korea Animal Improvement Association	Institutions: Grading evaluation of meat quality, feedback of breeding sow ability examination results
Local government	Boryeong and Hongseong in Chungcheongnam province	Administrative support and business unit management

Source: MAFRA (2013b)

June 2010. The Cluster started out with 1,800 participating farmers, and the number of participating farmers expanded to 2,800 by 2012. The capital status for 2012 was 3.3 billion KRW, invested by 432 farms. The national budget of 3.1 billion KRW was invested from 2009 to 2012.

Table 14.5 below shows the cooperative system between industry and academia for the Persimmon Cluster. Farms harvest persimmons, companies manufacture processed produce (e.g. dried persimmons, persimmon extracts, vinegar, and wine), and universities develop feed and new products. Manufactured products include dried persimmons, astringency-removed persimmons, persimmon extracts, and tannin, among others. In addition, five persimmon-harvest experiential orchards and direct outlets (e.g. Korea Train Express (KTX) station, highway rest-areas, and department stores) were operational. Efforts were also made in foreign market development and resulted in 150,000 USD of exported persimmon products to countries such as Japan, the United States, and Canada in 2009.

Revenues expanded 7.4-fold over the course of 3 years of operation from 700 million KRW in 2009 to 5.2 billion KRW in 2012. To attain higher added value for persimmons, various products were developed using persimmon by-products, off-standard persimmons, and persimmon leaves. Currently, 20 products are being produced.

3.2.3 The Melon Cluster Case

Gokseong Melon Cluster, located in the Gokseong region in Jeollanam-do Province, was established in January 2009 as an agricultural corporation, and 206 farms participate. The invested capital is 12.9 billion KRW, and all participating farms have contributed. A national budget of 6.2 billion KRW was invested from 2008 to 2012.

Table 14.5 Persimmon cluster industry–academia cooperation system

Category	Participant companies and institutions	Role
Industries	Farms : 5 farm organizations, about 2,800 farmers	Farms: Cultivate and produce persimmon, persimmon processing
	Companies: MSC, Persimmon Wine, Sangju F&G Agricultural Associations and seven other companies	Companies: Produce and sell persimmon extracts, concentrate, persimmon wine, and dried persimmon, etc.
Academies	Universities : Yongnam University, Suwon University, Kyongbuk Science College	Universities: Develop high consumption products, extraction and concentration technology development, functionality certification, product development
	Institutions: Korea Food Research Institute, Agricultural Technology Center, Rural Development Administration	Institutions: Develop processing standards, provide education
Local government	Cheongdo, Sangju, and Mungyeong in Gyeongsangbuk-do province	Administrative support and business unit management

Source: MAFRA (2013b)

Table 14.6 Melon cluster industry–academia cooperation system

Category	Participant companies and institutions	Role
Industries	Farms: 206 farms Companies: Melon Breeding Research Center, Gokseong Seedling Raising Co., Gokseong Logistics, Honam Shani Co. Ltd	Melon Breeding Research Center: Research and supply high quality melon varieties Gokseong Seedling Raising Co.: Raise and provide melon seedlings Honam Shani Co.: Manufacture processed products
Academies	Universities: Sunchon National University, Chonnam National University	Universities: Develop responsive technology for continuous cropping hazards, marketing and sales methods
	Institutions: Biological Control Center, Jeollanam-do Food Research Center	Institutions: Research insect damage, develop processed products
Local government	Gokseong in Jeollanam-do province	Administrative support and business unit management

Source: MAFRA (2013b)

Table 14.6 below shows the cooperative system between industry and academia of the Melon Cluster. Farms are responsible for cultivating melons. Companies are responsible for growing seedlings and processing. Universities and institutions are responsible for responding to continuous cropping hazards, marketing and sales methods, insect damage research, and processed product development. Manufactured products include melons, melon rice snacks, and melon vitamin candies, among others. Additionally, to attract green tourism clients, the cluster cooperates with Gokseong Train Village to promote tastings, excursions, and advertisements.

Revenues increased 2.9-fold from 1.4 billion KRW in 2008 to 4 billion KRW in 2012. Exportation efforts have resulted in selling 80 tons of melons, worth 300 million KRW in 2012, to Japan and Singapore. When the participation effects for each entity are observed, a farm's average production increased as a result of melon-cultivation technology education, companies bred high-quality melons and distributed specimens to standardize varieties, and universities and institutions developed various processed products through cooperative research.

4 Planning Food Industry Clusters in Gangwon-Do Province

4.1 Various Types of Food Industry Clusters

In the future, regional food industry clusters will gradually expand. Therefore, various differentiated clusters should be established with unique items and functional ingredients based on the regional characteristics. From the simple item-based processing clusters of today, diversifications such as high-level processed food

14 Promotion Policies for Food Industry Cluster in Korea   191

development, function-centric clusters, and traditional food (ethnic food) and local food clusters are necessary.

With this in mind, food industry clusters can be divided into the four types below, which suggest clusters unique to Gangwon-do Province. Of course, combining multiple types is possible according to the area's conditions.

4.1.1 Types of Food Industry Clusters

Item-Based Clusters

In this cluster we can develop many processed products using largely produced items such as highland vegetables, potatoes, corn, beans, rice, minor grains, apples, peaches, grapes and herbs from Gangwon-do Province. We can also enhance the added value by connecting up with direct marketing and experiential tourism. Figure 14.3 below shows several examples of this type, for example, the Herb Cluster, Korean Beef Cluster, and the Wine and Traditional Liquor Cluster.

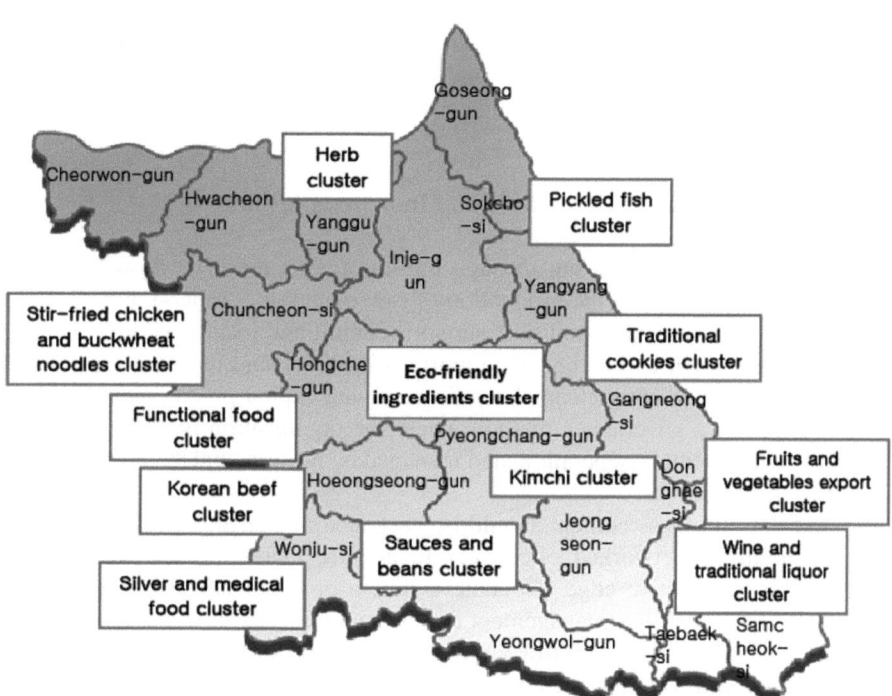

Fig. 14.3 Various types of food industry clusters in Gangwon-do province

Functional Clusters

This type of cluster is focused on the functional aspect, such as creating functional food of different ingredients specifically targeting the elderly, patients, children and other high-demand customers. This type includes a Functional Food Cluster, a Silver and Medical Food Cluster, an Eco-friendly Ingredients Cluster, and a Fruit and Vegetable Export Cluster.

Traditional Food Cluster

This cluster primarily focuses on the development and selling of old Korean ethnic foods made with traditional ingredients and methodologies. This includes, for example, a Kimchi Cluster, a Sauce and Bean Cluster, and a Traditional Cookie Cluster.

Local Food Cluster

Stir-fried chicken (*dak-galbi*) and buckwheat noodles (*mak-guksu*) are nationwide famous local foods of Chuncheon City in Gangwon-do Province. We can make a Stir-Fried Chicken and Buckwheat Noodle Cluster in this city. This specific type of cluster is developed based on the special local food as in the above example. In this case, it is desirable to connect local food with the forward- and backward-related industries.

4.1.2 The Governance Method of Food Industry Clusters

Cluster projects are carried out through the correlation among various agencies within the cluster, such as the production of raw materials (farmers), organization (Agricultural Cooperatives), political support (central and local government), new product development and technical support (universities and research facilities in the region), and marketing channel acquirement and capital procurement (companies). Therefore, a governance method where the organization is formed through civil and governmental cooperation, and managed by professional management personnel, is most desirable.

In addition, business content should be packaged and agencies involved in the cluster should be networked effectively to acquire marketing channels and to improve the competitive edge. To achieve this, a project leader is necessary for organically fusing various management entities and items to allow business to proceed smoothly. From this point of view, it is also very important to nurture good regional leaders who understand not only regional features and business content, but management knowhow.

4.2 Correlation Between the Food Industry Clusters and Related Institutions

4.2.1 The Relationship Among Food, Tourism, and Region

Figure 14.4 below shows the intimate relationships among food, tourism, and region. In the case of Gangwon-do Province, the tourism industry is highly developed, and through the Rural Community Renewal Campaign, regional development is highly active. If these factors are well-connected, a large amount of development is expected within the food industry.

Before this can happen, a connection with the tourism industry to stimulate food tourism is required. When food and region are connected, the concept of a local food system is created. The Japanese local consumption of local products and Europe's slow food movement are examples of this. Some specific cases of interconnecting rural regions with tourism are green tourism, urban–rural exchanges, the "Tying up Campaign of One Company with One Village", and the Rural Community Renewal Campaign of Gangwon-do Province.

4.2.2 The Correlation Between the Traditional and Local Food Clusters and Related Institutions

Traditional and local food reflects a region's character and is highly popular among tourists. Recently, there are many experiential events where tourists make traditional food by themselves. Small-scale traditional and local food clusters in the

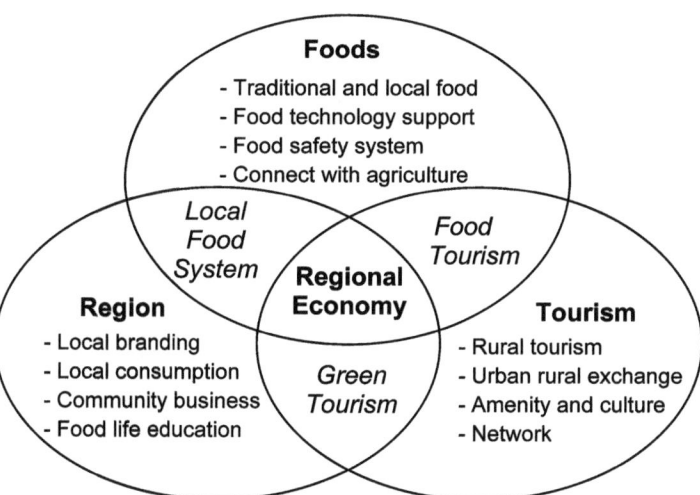

Fig. 14.4 The relationship among food, tourism, and region

region have a low output and weak marketing power, which is why direct sales to tourists is most efficient.

However, in many cases, traditional and local food has low-quality equalization due to lack of standardized recipes. Additionally, new product development efforts are necessary to meet the tastes of children and young consumers while maintaining the characteristics of traditional and local food.

Figure 14.5 below shows the correlation between the Traditional and Local Food Clusters and related institutions. First of all, it is necessary to establish the Food Development and Technical Support Center (FDTSC) within the Gangwon-do Province Agricultural Research and Extension Services (GARES) to proceed with technical support and safety evaluation.

In addition we can use the Gangwon-do Province Agriculture Produce Genuine Center (GAPGC) located in Seoul and run by Gangwon-do Province as an antenna shop, to observe consumer responses to the newly-developed products of the Traditional and Local Food Clusters, and for advertising operations. Consumer trends should also be analyzed beforehand to reflect on new product development.

An interconnection between food industries and tourism programs within the region must be well-established to inform tourists in their selection of tour programs. By carefully considering tourists' desire for traditional and local food information (e.g. traditional and local food by season and prices), the efficacy of budgets for potential cuisine and setting can be better comprehended by the tourists themselves.

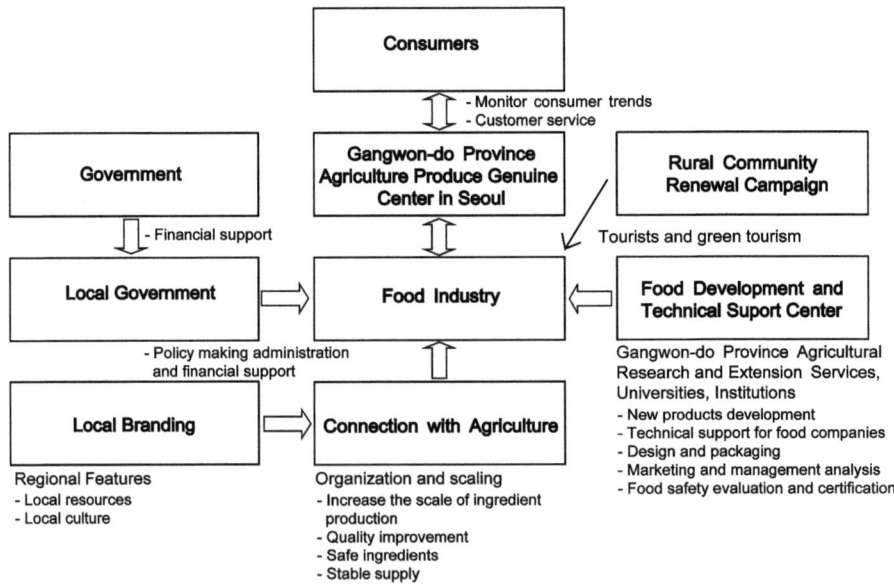

Fig. 14.5 Correlation between food industry clusters and related institutions

5 Conclusion

The South Korean government has been developing a National Food Cluster in Iksan City, Jeollabuk-do Province since 2008 to lead the development of regional small scale food industries, and to expand R&D capabilities within the food industry. In this complex, a Food R&D Research Center and other supportive facilities will be present and are expected to serve as a hub in the development of the South Korean food industry clusters.

In addition, South Korea has been implementing food policies, including the Regional Strategic Food Industry Promotion Policy since 2005, to promote food industry clusters connected with agriculture in the region. In 2014, food industry clusters are operating in 77 localities throughout the country, and this will expand to 100 localities by 2019. Recently food industry clusters have expanded their business areas into several city and county coalitions, and also produce various and high-quality products in comparison with the initial stage. However, many cluster business units still have several common problems to improve, such as insufficient experts, weak self-reliance, and lack of ability in business model development and marketing activities.

To develop the food industry in Gangwon-do Province, various differentiated clusters should be implemented with unique items and functional ingredients based on the regional characteristics. Four types of clusters were suggested, such as item-based clusters (e.g. herb, and Korean beef clusters), functional clusters (e.g. functional food, and silver and medical food clusters), traditional food clusters (e.g. kimchi, sauce and bean clusters), and local food clusters (e.g. the Stir-Fried Chicken and Buckwheat Noodle Cluster).

Also, by interconnecting a vast tourism infrastructure and bringing tourists into local food industries, the food industry range might be expanded and produce a synergistic effect from mutual complementary relationships between tourism and food industries. For example, food tourism, the local food system, and green tourism functions may all be stimulated.

Traditional and local food should be differentiated by not only its quality and functionality, but also in design as well, and this requires advanced technical support. To achieve this, it is necessary to install a Food Development and Technical Support Center (FDTSC) within the Gangwon-do Province Agricultural Research and Extension Services. By cooperating with universities and research centers in the region, the FDTSC can support small-size food companies for various areas, such as new product development, technical support, designing and packaging, marketing and management analysis, and safety evaluation, etc. Also Gangwon-do Province Agriculture Produce Genuine Center (GAPGC), located in Seoul, can be used as an antenna shop, to observe consumer responses on the newly-developed products of the clusters, to monitor market trends, and for sales promotion.

References

Choi JH (2013) Food industry mid-long term improvement plan for creating new added value in agriculture. Korea Rural Economic Institute, Seoul

Lim JB (2012) Strategy for global food processing invigoration. Seoul National University R&DB Foundation, Seoul

MAFRA (2013a) Regional strategic food industry promotion

MAFRA (2013b) Five year plan for regional strategic food industry promotion project

MAFRA (2014) Improvement plans for regional strategic food industry promotion project

Chapter 15
The Trends and Potential for Food Industry Clusters in Korea

Jaehyeon Lee

Abstract So far, Korean rural areas could hardly have chances to arising food industry clusters, due to weak industrial infrastructure of non-agriculture sectors. However, for adjust to WTO Agriculture Agreement upon implementation, the government took the various measures in domestic agriculture policies be focused on the development of regional-agriculture cluster and branding of agricultural production. Followings are characteristics and problems concerning to agricultural policy and challenges be related to food industry cluster in Korea.

First, the agri-food policy has been actively involved to the branding of agricultural productions and development of agri-food cluster in Korea. The development of standards in the marketing of food distribution, branding guidance to agricultural products and pilot projects for the food cluster formation, are the related efforts. Secondly, the most noticeable pattern is to expand the business area as like production line expansion and finding service or processing sectors in local level agri-food cluster. Thirdly, many factors prevent the development of the cluster be required cooperation as well as competition among related local companies. To point out problems, it could be summarized spot transaction on products sales, exclusive in-sourcing led by production side and dependence on government subsidies.

Keywords Korea • Food industry clusters • Branding of agricultural products • Expand the business area • Competition and cooperation

1 A Perspective of "Food Industry Cluster"

1.1 The Role of Local Brands in Clusters

The concept of clusters as a source of competitiveness was initiated by Michael E. Porter and evolved into the Cluster Initiative Model (Sölvell et al. 2003), detailing the keys to success, including what kind of organized approach is desirable and

---

J. Lee (✉)
Faculty of Agriculture, Kagoshima University, Kagoshima, Japan
e-mail: lee@agri.kagoshima-u.ac.jp

what process would lead to success. The concept is used widely across the world in national industrial policy or rural development initiatives.

According to the Cluster Initiative Model, building local brands is an effective marketing tool if market information is organized and shared. There is a general consensus that rise in brand recognition helps the cluster grow further because higher recognition draws related businesses and human resources into the cluster (Sölvell et al. 2003, p. 10).

1.2 The Definition of a Local Brand

A **local brand** can be defined as "a brand used in a unified manner to differentiate, from others, the goods and services related in some way or other to the industry, history or natural environment that characterizes the locality, or to the image/symbol of the locality generated by its industry, history or natural environment."

In order to link the local brand to an industrial cluster, it is first necessary to put in place a set of uniform standards that covers the processes from production to marketing and to have them applied to the producers and products within the region. Then it is important to build a **brand hierarchy** where **category and product line are expanded** to bring the brand power into full play in the market (Lee 2010, p. 3).

1.3 Competitiveness of Food Industry Cluster

Competitiveness of the food industry cluster at the farming community level seems to come from saving **transaction cost, reduced risk**, heighten **added value** and **expansion of business domain**. Transaction cost decreases less in a food industry cluster because the supply chain and value chain of the production and sales of foods including farm products is contained in a certain geographical area. It is considered appropriate to expand the current business into the processing industry for further enhancement of the business domain.

2 Korea's Approach to Food Industry Cluster

2.1 Food Industries Located in Rural Areas

In Korea, the percentage of farmers who work part time (37.4 % in 2005) and the percentage of their income coming from side jobs (10.7 % in 2009) illustrate how few opportunities there are for them to find work other than farming. In rural areas left outside the large clusters of the industrial estates built after the war, there are very few related industries and little buildup of the human resources necessary to form a food industry cluster. When we take a look at the region-by-region share of food product sales, 43.8 % are in the Seoul Capital Area. If the Chungcheong region is included, the share rises to as high as 67.5 % (Korea Ministry of Agriculture 2007).

The government chooses the location of industrial complexes and there are few opportunities to launch food-related businesses. The business environment in rural areas is a typical example of "the old structure led by local government and cooperatives" – where competition does not exist between producers and there is no driver for innovation – and is the direct opposite of the cluster concept advocated by Michael Porter (1998, p. 79). However, there are signs of change in the old structure as domestic agriculture is being liberated as a result of the implementation of the WTO Agreement on Agriculture.

2.2 WTO Agreement Measures

Korea put in place a variety of different agricultural measures domestically in order to prepare for the implementation of the WTO Agreement on Agriculture.

Regarding these, the government has been proactively implementing various support measures: the establishment of marketing standards (the standardization of quality, specifications, and safety, and systems for labeling and certification) to differentiate domestic farm products; the introduction of farm-product collection and delivery facilities (APCs: Agricultural Products Processing Centers; and RPCs: Rice Processing Complexes, etc.) to support marketing efforts for the producing area; and support for brand-building efforts for specialty and other farm products. In particular, the labeling and certification systems for farm and food products have a wide coverage, ranging from traditional foods to functional foods. New systems are being put in place, unnecessary ones are being abolished and necessary revisions are being made to the existing ones in a speedy manner (Lee 2008a).

2.3 Local Agriculture Cluster Model Project

Local Agriculture Cluster Model Projects are a policy tool of local industrial revitalization that integrates rural communities and industrial policies. A local agriculture cluster is defined as "a concentration of agricultural industry to cause innovation in local agriculture by utilizing the local resources in an optimum manner through a network of business–government–academic communities and players involved in the production, distribution and processing of farm products associated specifically with the locality" (Ministry of Agriculture 2008). According to the concept for the model, local producers, related support organizations, research and development institutions, and financial institutions all work together to build a cluster in an organized fashion (Ko 2008).

The actual model project cases (20 localities) have in place a promotion system mobilizing producers' groups, universities and research institutions, and local businesses, and are working on marketing for the producing area – quality improvement and channel management – and the establishment of processing businesses (Table 15.1).

In addition, many of them are utilizing local brands (Table 15.1) which are worthy of note. Incidentally, the geographical indication system (2009) has given 65 items the right to use a geographical name exclusively as a trademark.

2.4 Food Industry and Korea National Food Cluster

In 2008, the Ministry of Agriculture changed its name to the Ministry for Food, Agriculture, Forestry, and Fisheries, and the "Food Industry Development Plan" was announced immediately after 2008. Amongst other measures, food-related industries have high expectations for the "Korea National Food Cluster (2009–2015)".

The Korea National Food Cluster estimates the global food market to be worth US$4 trillion (Fig. 15.1). One trillion won will be invested to develop a 400-ha industrial park where food export businesses will be invited to set up operations. Research and development and other supporting institutions will be built or relocated here to lay the foundations for networking and innovation.

Reference is necessary to the challenges for functional food development. There are efforts to develop functional food and foodstuffs for functional food in Korea. A certification system for functional food is already in place. On another front there is the marketing drive to spread Korean food worldwide. Target foods include traditional fermented foods, traditional local cooking and traditional wine.

3 Business Expansion Caused by Development of Local Brands: Three Cases Studies

The following case studies illustrate how agriculture-related businesses have been generated in the localities where farm products were developed into a local brand with a certain level of recognition.

3.1 Development of Food Processing led by Agricultural Cooperative

Hamyang Agricultural Cooperative is a large-scale regional agricultural cooperative in Hamyang County, Gyeongsangnam-do [South Gyeongsang Province], with 5,364 full members and 4,822 associate members. The cooperative made an early move in response to the branding policy, obtaining certification and brand

15 The Trends and Potential for Food Industry Clusters in Korea

Table 15.1 Categories of local agriculture clusters

Type	Type of innovation leader			Business type			
	University/ institution	Producer group	Related business	Local government	Production/distribution	Processing	Theme
City/country area							
Single city/ county	Yeongdong (grape) Punggi (ginseng)	Anseong (cattle, Korean native cattle, fruits/ berries, perishables, eco-friendly items, etc.)	Pcheon (Korean sweets)	Boseong (green tea) Goesan (red peppers) Asan (organic farm products) Seocheon (flax) Jangsu (apples) Jeongeup (circular agriculture) Imsil (cheese) Hampyong (scientific farming) Hadong (green tea)	Anseong (many) Goesan (red peppers) Jangsu (apples) Imsil (cheese) Boseong (green tea) Punggi (Ginseng)	Pocheon (Korean sweets) Yeongdong (grapes) Seocheon (flax) Hampyong (scientific farming) Hadong (green tea)	Asan (organic farm products) Jeongeup (circular agriculture)
Multi city/ county	South Gyeongsang (pig farming)	Gangwon (Korean native cattle) Yeongwol (pork) South Jeolla (rice) South Gyeongsang (rice)			Gangwon (Korean native cattle) Yeongwol (pork) South Jeolla (rice) South Gyeongsang (rice)		
Total	3	5	1	9	11	5	2
Greater area	North Gyeongsang (Korean native cattle)			Jeju (Citrus fruits)	North Gyeongsang (Korean native cattle) Jeju (citrus fruits)		
Total (20)	4	5	1	10	13	5	2

Source: Ministry of Trade, Industry and Energy, and the Presidential Committee on Balanced National Development (2007), p. 158

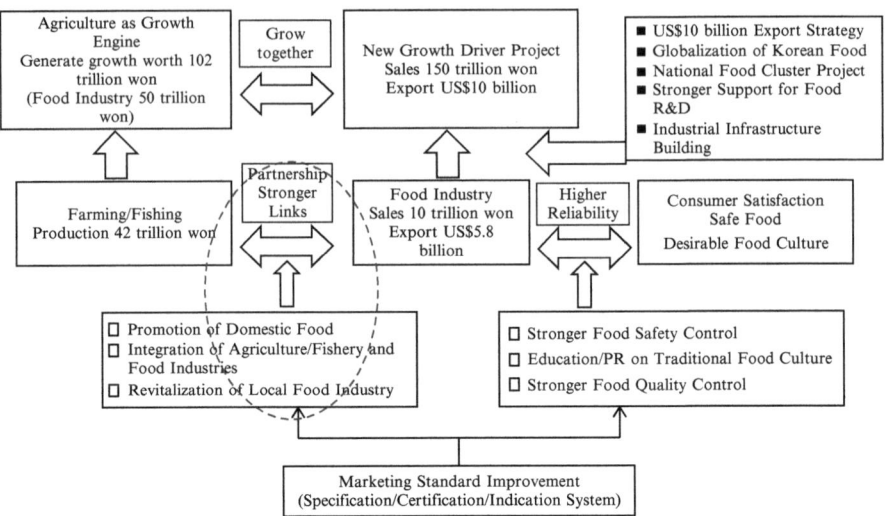

Fig. 15.1 System of Korea's comprehensive food development measures (Source: Ministry of Agriculture, Food and Rural Affairs 2008, p. 42)

development (Saito 2008, p. 283). The area lies at the foot of Jirisan – one of the most important mountains in the country – and boasts many blessings of nature such as clean water and a beautiful forest landscape.

These resources went into the development of a packaged rice brand: "Jirisan Hwang-to", i.e., "Mount Jiri loess" in English. As there were few other packaged rice products at the time with quality certification and brand value, some large retailers requested the right to sell it exclusively. Thus, the Hamyang Agricultural Cooperative agreed to supply most of its packaged rice products directly to these retailers in large quantities. Private brand products account for about 40 % of the total sales of volume retailers. Mutual trust was built up between the cooperative and the retailers. What happened subsequently was that the sales of minor grain products rose gradually as they were placed on sale next to the packaged rice section. A new farm product processing facility was built in addition to the RPC (Rice Processing Complex). The new facility produces and sells not only minor grains which are a derivative category of rice and powdered grain products, but also functional food that one readily associates with minor grains or powdered grain (Fig. 15.2).

Branding of a rice product gave Hamyang Agricultural Cooperative an opportunity to develop a retail channel and to supply private brand (PB) products to large retailers. Cooperation on the retailer's marketing strategy led to a new product lineup and sales of new products in an adjacent sales-space. This is an example of a successful case of expanding business from supplying fresh farm products to the processing of a number of different products.

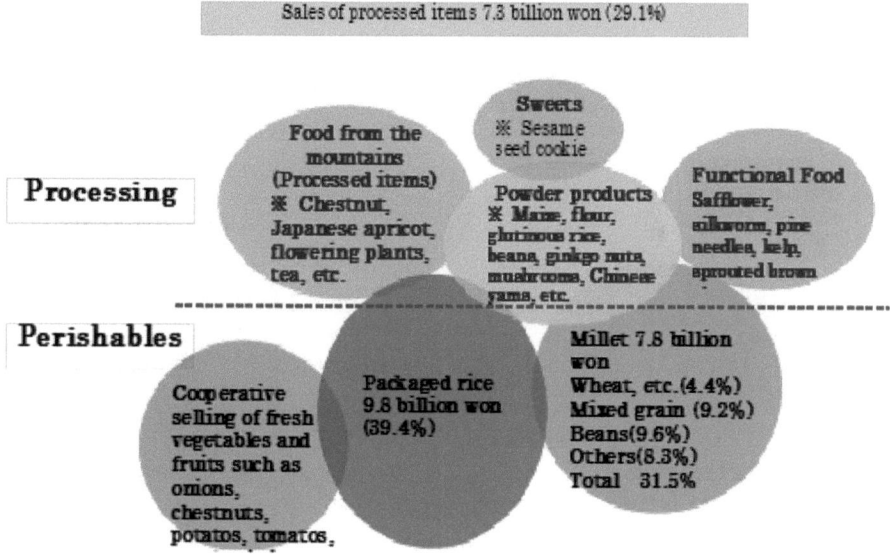

Fig. 15.2 Hamyang Agricultural Cooperative's sales by categories

3.2 Management of Local-Government Brands and Local Brands

Buyeo County in Chungcheongnam-do [South Chungcheong Province] was the former capital of Baekje, an ancient kingdom of Korea. The county developed the GOODTRAE brand in 2003 and acquired a license for it. First, out of the farm products grown in the county, the brand designated eight items which had a relatively high production share. The eight were labeled as the **Eight Delicacies of Buyeo** (mushrooms, cherry tomatoes, watermelons, melons, shiitake mushrooms, chestnuts, cucumbers and strawberries). The suppliers of these items were required to meet a set of quality standards and supply conditions before they were permitted to use the brand name. As of 2008, 59 suppliers are licensed to use the brand. There are 3,200 farmers involved, and 70 % of deliveries are labeled with the brand. The sales of branded products amount to approximately 75 billion won (the eight items only).

The local government used the subsidized project to develop large-scale promotional activities and gave the producers a thoroughgoing education on quality control. As a result, the brand has won many awards relating to brand evaluation. Recognition rose gradually thanks to the media and an increase in the number of clients, such as large retailers.

Encouraged by the high recognition, the local brand GOODTRAE of Buyeo County expanded the brand category after it had built brand recognition for the eight

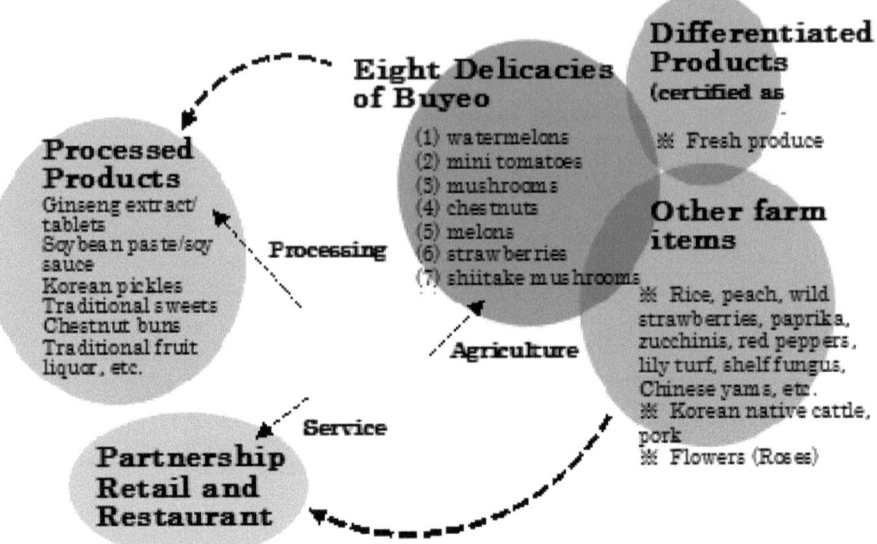

Fig. 15.3 The picture for GOODTRAE category expansion

farm products. Currently, five categories use the GOODTRAE brand: (1) the Eight Delicacies of Buyeo; (2) other fresh farm products produced locally; (3) organic farm products; (4) processed farm products; and (5) restaurants and retail outlets that procure GOODTRAE brand products (Fig. 15.3).

3.3 Networking and Innovation in Clusters

Here, the Traditional Medicine Cluster development project is introduced as an example. It involves a wide selection of players such as producers of medicinal herbs, testing laboratories, pharmaceutical companies, traditional medicine hospitals and producers of functional food in two administrative jurisdictions (Daegu and Gyeongsangbuk-do [North Gyeongsang Province]) (Lee 2008b). The rugged mountain terrain of Gyeongsangbuk-do has famously been utilized to grow apple and other fruit trees for a long time. In addition to fruit trees, many farmers in the region have been growing medicinal herbs as industrial crops on mountain farms. As there are few areas in the country that meet the climatic and soil conditions to grow medicinal herbs, the crop has become the region's specialty. Actually, many of the items for which the region has acquired a geographical indication license are medicinal herbs.

As consumers place greater importance on health and safety in choosing food items, recently they have been showing a preference for functional food. Also, traditional medicine is part of ordinary medical practice in Korea. These factors are behind the growing demand for health foods and traditional medicine. The "Traditional Medicine Cluster" development project is a local industrial policy meant to develop the medicinal herb industry, as the market is growing and demand conditions are met.

The Traditional Medicine Cluster has the Korea Promotion Institute for Traditional Medicine Industry at its core body. It will attract pharmaceutical venture businesses to the site, and build a network of research and development institutions – consisting of universities and research institutions – and the producing area for medicinal herbs so as to share the knowledge and information necessary for the development and marketing of new products.

The Traditional Medicine Cluster project is remarkable in a number of ways. First, the Korea Promotion Institute for Traditional Medicine Industry acts as a vision provider. It runs a portal site that supplies a broad range of information about traditional medicine and acts as a matchmaker between research and development institutions, such as universities, and related businesses (Fig. 15.4).

Second, it places great importance not only on developing and spreading new varieties of herbs in the producing area, but also building marketing standards (standard specification of delivery, internal quality standards, and certificate/indication systems, etc.) for the raw materials (traditional herbs) as an important support measure for the traditional medicine industry. Third, in addition to linking traditional medicine and medical industries, the cluster project makes the best use of being a producing area of medicinal herbs to generate a broad range of service industries, such as the health food industry (honey, traditional functional tea, and traditional herb drink products, etc.), the health and sports industry (fasting or meditation facilities, and tai chi chuan training studios, etc.), and the traditional medicine leisure industry (traditional medicine saunas, natural forest land, and shiatsu, etc.) (Fig. 15.5).

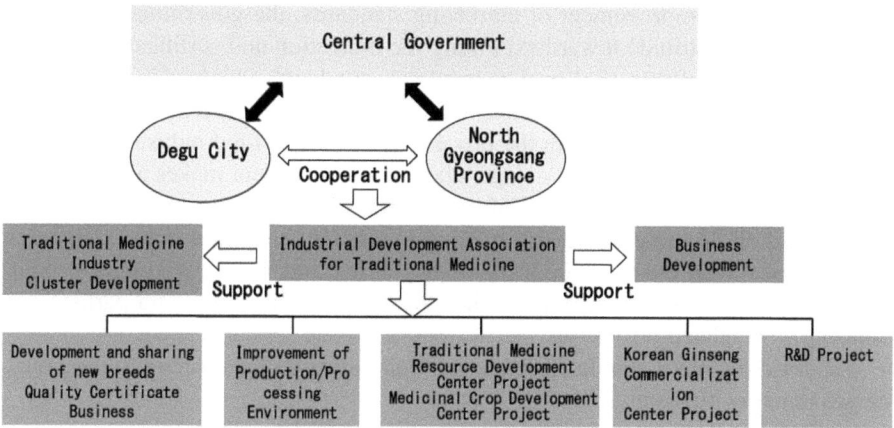

Fig. 15.4 Support structure for traditional medicine cluster

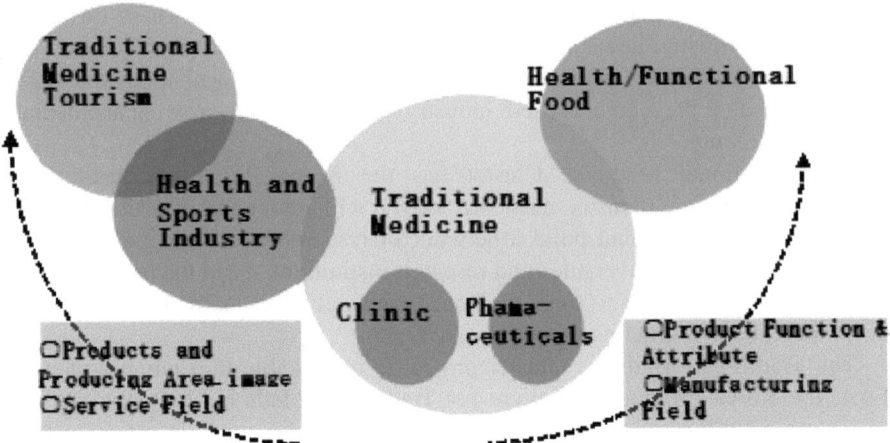

Fig. 15.5 Expected business fields of traditional medicine cluster

4 Features of and Challenges for Korea's Food Industry Cluster Approach

4.1 Features

The first feature concerns the promotion system. The government has a promotion system for the development of a food industry cluster and branding which is characterized by the following three points. (1) The government has a program to develop a food industry cluster systematically. The policy is implemented in a coordinated manner at the national and local government levels. (2) As rapid progress was made in the improvement of marketing standards, the government has maintained a positive attitude toward exploiting the indication and certification systems in brand-building efforts. (3) Local governments are deeply involved in the development and management of local brands and in forming local industry clusters.

The second feature is seen in the process of forming a food industry cluster. At the moment, it is still at an early stage where the government makes approaches to localities to improve the necessary infrastructure. Another noteworthy feature is that brand-building is going ahead and gaining a positive reputation from consumers, which comes from cluster-wide technology sharing and harmonization of product quality at a high level as a result of the clustering of businesses and partnership-building with related businesses.

The third feature is that in Korea, local food industry clusters tend to have businesses in many different fields.

4.2 Challenges

There lie a number of challenges in the food industry cluster initiative strategy and management of local brands.

First, there is no systematic brand management system in place to control, under relevant quality and safety standards, the many different items supplied by the many producers in the region.

Second, whether fresh produce or raw material, deals tend to be spot transactions to supply products through a middleman or a vendor.

Third, if the operation is taken care of by a management body, particularly by the local agricultural cooperative, the possibility is high that the relevant businesses will have few opportunities to participate.

Fourth, a close look at any food industry cluster project reveals a construction-oriented business structure that relies heavily on government subsidy.

We need to recognize that reliance on spot transactions in deals, in-sourcing by brand-management organizations and the government-led approach may each act as a roadblock, as clustering requires competition and cooperation among related businesses.

References

Ko B (2008) Development of local agriculture cluster in Korea. In: Yagi H (ed) Economic interdependence and agriculture in North-East Asia – competition and cooperation under the formation of regional economic zone. University of Tokyo Press, pp 205–222, Tokyo [in Japanese]

Korea Ministry of Agriculture (2007) Food industry general statistics, Seoul [in Korean]

Korea Ministry of Agriculture (2008) Local agriculture cluster revitalization support, Seoul [in Korean]

Lee J (2008a) Problems on branding of agricultural products as a domestic market-opening measure. In: Yagi H (ed) Economic interdependence and agriculture in North-East Asia – competition and cooperation under the formation of regional economic zone. University of Tokyo Press, pp 126–140, Tokyo [in Japanese]

Lee S (2008b) Degu and North Gyeongsang Prefecture traditional medicine cluster development plan. Degu and North Gyeongsang Research Institute, Degushi [in Korean]

Lee J (2010) Product line and brand hierarchy of packaged rice in Korea – case study of agricultural cooperative's RPC. J Food Syst Res 16(4):1–13, Tokyo [in Japanese]

Ministry of Agriculture, Food and Rural Affairs (2008) Food industry development plan, Seoul [in Korean]

Ministry of Trade, Industry and Energy, the Presidential Committee on Balanced National Development (2007) Annual report on balanced national development plan, Seoul [in Korean]

Porter ME (1998) Clusters and the new economics of competition. Harv Bus Rev Nov-Dec:77–90, Boston [in English]

Saito O (ed) (2008) Strategy and management of local brands – Japan and Korea/from rice to sea food. Rural Culture Association Japan, Tokyo [in Japanese]

Sölvell Ö, Lidqvist G, Ketels C (2003) The cluster initiative green book. Center for Strategy and Competitiveness, Stockholm [in English]

Chapter 16
The Promotion of and Challenges for the Agricultural Senary Industrialization Policy in the Republic of Korea

Youkyung Lee

Abstract In this paper, I have summarized the background to the promotion of the agricultural senary industrialization policy in the Republic of Korea (ROK) and the trends in related policies to date, and have considered the challenges for the future. Senary industrialization has commonality with industrial clusters in the aspects of integrating and linking production and each level of entity in processing, distribution, sales, and research and development, creating value, and developing a synergistic effect. Accompanying the progress of internationalization of the agricultural produce market in the ROK, the rise in agricultural incomes and the securing of places of employment, etc., are being hastened. Within the senary industrialization policy, which is one of the measures to tackle this, there are cases which take project entities which are composed of multiple entities as policy targets, and cases which integrate and connect the production processes for agricultural produce and food items within a division of labor, forming one kind of cluster, and attempting value creation. In the current situation there are many networks connected by mere "lines". However, the related entities deepening the reciprocal exchange of information and forming relationships of trust is an essential factor for the formation and endurance of clusters. In the future, it will be necessary to devise initiatives for making possible the construction of sustained linked relationships among the participating entities.

Keywords Senary industrialization • Cluster • Linkage • Value creation

---

Y. Lee (✉)
Department of International Development Studies, College of Bioresource Sciences Nihon University, Fujisawa, Japan
e-mail: lee.youkyung@nihon-u.ac.jp

1 Introduction

Michael Porter, the propounder of the cluster theory, held that "a cluster is a geographically proximate group of interconnected companies and associated institutions in a particular field", and that the entities within the cluster are connected by commonality and complementarity (Porter 1998). That is, a cluster may be said to be a group in which various entities which are operating within a fixed regional ambit are taking advantage of their mutual strength, and along with heightening its productivity while compensating for the areas which are lacking, creates new value. Moreover, as the enduring elements of clusters there are innovation, collaboration, knowledge creation, collective study, and the cultivation of human resources, and it is held that within the cluster, with the various entities creating new value via a collective study process, the cluster will endure and evolve (Futagami and Hioki 2008).

Meanwhile, what is the senary industrialization of agriculture? The senary industrialization in the ROK's agricultural policy has been defined as "the activity to create new added value by the linking up of agriculture based on all the tangible and intangible resources which exist in agricultural communities and the manufacturing and processing of food and specialty goods (secondary industry) and distribution and sales, and cultural, experiential, and tourist services (tertiary industry)". That is, if there is the policy support to facilitate initiatives on the activities for creating new added value, utilizing the various resources present in agriculture and rural areas, and also support toward initiatives for individual operations, the policy of the senary industrialization of agriculture will also include policy projects to promote the formation of one kind of cluster linking up the various related regional entities, such as the producers of agricultural products within the area, processors and distributors, sales, service providers, and additionally research and public institutions.

With the agricultural senary industrialization policy having been worked out from the fact that the production and processing, distribution, sales, and research and development entities to date have adopted their respective divisions of labor separately, it can be said that the production processes for agricultural products and food in the ROK have commonality with industrial clusters in the aspects of the integration and linking up of each level of entity, the creation of new value, and the development of a synergistic effect.

Next, in the agricultural policy of the current administration (President Park Geun-hye, from February 2013) which brought up "creative economy" as a national policy slogan, they raised the "activation of the agricultural senary industrialization policy" in which the production of agricultural products, and processing, distribution and tourism, etc., were integrated, and as specific policy projects thereof, there are such things as: the nurturing of distinctive local industry and the putting in place of joint processing centers for agricultural and livestock products; support for firms which have linked up agriculture, commerce, and manufacturing; and the activation projects for rural tourism via urban–rural exchange. Although the using of the term "agricultural senary industrialization" among the policy projects within the ROK's agricultural policy was a first for the current administration, the off-farm income policy and rural-area revitalization policy, among others, which included the same concepts, have been carried out as before.

Consequently, in this next section, I will summarize why the agricultural senary industrialization policy in today's ROK is the focus of attention, giving the background to its promotion, and the current state of play for the related senary industrialization policies which have come into force to date, and consider the challenges for the future.

2 The Background for the Promotion of Senary Industrialization Policy in the ROK

2.1 The Farmers' Income Issue

In ROK agriculture, full-time farmers account for more than 50 % of the total. It would be no exaggeration to say that at the core of the problems for agriculture and agricultural communities is the issue of farm income. According to the results of research by Statistics Korea, in 2012, with farmers' income approximately 31 million won, and that for urban working households approximately 54 million won, the situation became one where farmers' income was no more than 57 % of that for urban working households, and the income disparity between the two groups was great (Lee 2013).

Incidentally, related to agricultural income, the amount of agricultural production for 2011 had increased 15 % compared to 1995. Meanwhile, the unit production of agricultural produce and the amount for labor inputs decreased by 13 %, and the real value added increased 20 %. In the period from 1995 to 2010, however, consumer prices rose 72 %, and the price for agricultural material inputs rose 126 %, whereas the rise in the price of agricultural produce stalled at 28 %. That is, although agricultural production increased and efficiency heightened, it is considered that agricultural income decreased as the rate of increase in agricultural production costs was far lower than the rate of increase in the cost of material inputs and consumer prices. Moreover, the reason why the rate of increase in the price for agricultural products was low was the result of the volume of agricultural product imports having increased 80 %, and in addition domestic production rose 15 % (Lee 2012). Considering that the FTAs concluded with 47 nations to date will further be brought into effect in the future, because it is forecast that low-price agricultural products from overseas will increase yet further, a rise in price of domestic agricultural products is unlikely. Consequently, it can be said that measures for increasing income from sectors other than agricultural production are pressing issues.

2.2 The Idle Labor Force Issue

Heretofore in rural areas of the ROK the issue of idle labor had been highlighted. According to the 2012 statistical survey results, the annual farming management hours per farming household was 1,090 h, and the number of persons per farmstead was 2.55. That is, taking into consideration that while the farming management hours per month per farm worker can be calculated as 35.6 h, the average work hours per month in other industries were 180 h, the farming management hours were slightly less than one-fifth of the latter, and the idle labor force in the agricultural sector appears obvious. Furthermore, these labor hours have a strong seasonality, the disparity between the agricultural off-season from November to February and the busy farming season from May to October is large. Consequently, in agriculture and rural communities places for employment able to utilize effectively such seasonally idle labor are required (Lee 2013).

2.3 The Promotion of the Off-Farm Income Source Development Policy

The policy which was set out as a method for the acquisition of the above-mentioned off-farm income and the effective utilization of the idle labor force was the Off-Farm Income Source Development Policy, and has been carried out since the 1960s. Initially, it encouraged agriculture in a form with secondary jobs, such as domestic handicrafts, fruit trees, the growing of medicinal herbs and beekeeping, utilizing the idle labor force of rural areas which existed in large measure, and undertook provision of business finance and technological and management guidance to those undertaking domestic secondary jobs in farmers' groups. Subsequently, there was the promotion of agricultural and industrial estate development projects via the "Rural Community Income Source Development Promotion Act" (1983), and the rural tourism policy was promoted from the late 1990s.

In recent years, the "Law on Support for Farmers' Off-Farm Income Activities" (2010) came into force. This law had the objective: "Via supporting off-farm income activities making use of various resources including the produce of farmers and others, to provide opportunities for the expansion of off-farm income, and strive for the balanced economic development of the people". The law gave a legal foundation for the establishment work of agricultural produce processing facilities, and of agricultural produce technology utilization centers to support education, technological development, and consulting, etc., and currently, for industry and business start-ups, has become able to support in policy terms farmers' agricultural processed products.

2.4 The Promotion of the Rural Development Policy Emphasizing Regional Governance

The administration of President Roh Moo-hyun, which commenced in 2003, brought up "regional innovation" as a slogan, and the construction of innovative systems was promoted, including: the strengthening of the independence in business promotion of local government[1]; cooperation and partnerships among bodies within regions; the strengthening of industry–academia–government–research networks and learning; and a bottom–up designing of project plans. Taking advantage of this, the economic vitalization policies of rural areas have changed greatly, and have become able to stress the following three things: (1) the introduction of bottom–up formulae in the implementation of policy projects; (2) the promotion of local government competition via public-offering formulae; and (3) local governance. In particular, via stressing local governance, policy projects being advanced from a bottom–up format, and through the cooperation and learning of bodies within a region, intrinsic development concepts by means of the spontaneity of local people have come to be widely incorporated.

2.5 The Increase in Management Entities Running Agriculture-Related Businesses

It is possible to get a direct grasp of trends for agricultural senary industrialization from the statistics for farms which are conducting business other than that related to farm production. With surveys regarding related business in farm censuses having been undertaken from 2005, prior to that there were not many farms which conducted related business, to the degree where a survey category could be established.

From Table 16.1 it can be seen that in recent years farms which are conducting farm-related business have increased. In addition, despite the number of farms having decreased over the 5 years, the actual number of farms conducting related business has increased by some 50,000. Even taking into account the fact that from 2010 it has become possible to include farms acting as "agricultural machine operation agents" within those conducting farm-related business, there has been an increase of approximately 27,000 farms.

Next, the commonest farm-related business in 2010 was "direct sales points and direct sales" (10.0 %), followed in turn by "agricultural machine operation agents"

---

[1] The system of local government in the modern-day ROK began after the Local Autonomy Act passed the National Assembly on 4 March 1994, and the fourth local elections were put into effect on 27 June 1995. That is, as yet only 18 years have passed since the ROK's local government system was put in place.

Table 16.1 Number of farms conducting farm-related business (2005–2010)

	Total number of farms	Farms conducting farm-related business	Direct sales points	Direct sales	Processing	Farm restaurants	Agricultural machine operation agents	Farm guest houses	Rural tourism
2005	1,272,908	99,879 (7.8 %)	88,290 (6.9 %)		6,503 (0.5 %)	5,174 (0.4 %)	–(1) (–)	3,278 (0.3 %)	736 (0.1 %)
2010	1,177,318	151,515 (12.9 %)	28,127 (2.4 %)	89,107 (7.6 %)	8,564 (0.7 %)	9,043 (0.8 %)	23,331 (2.0 %)	4,468 (0.4 %)	

Source: Census of agriculture, forestry and fisheries for both years, Statistics Korea

Notes: (1) As the survey category of "agricultural machine operation agents" was set up in 2010, no survey data exist for 2005. (2) In 2005 the survey categories of "direct sales points" and "direct sales" were not differentiated. (3) In 2010 "farm guest houses" and "rural tourism" were surveyed as the one category of "rural tourism", including weekend farms. (4) It does not fit the total of farms conducting farm-related business and total of each types because several entities are engaged in multiple business. Multiple responses were allowed in the survey

Units: Households, %

(2.0 %), "farm restaurants" (0.8 %), and "processing" (0.7 %). Compared with 2005, the increases which stood out in particular were "direct sales points and direct sales" (6.9 % → 10.0 %) and "farm restaurants" (0.4 % → 0.8 %). This would appear to be an effect[2] that has emerged from the active promotion of "rural tourism".

Moreover, besides individual farms, within the business entities of agricultural corporations, business entities running non-farm-production business such as processing and distribution increased greatly from 1,941 corporations (58 %) in 2000 to 9,924 in 2013, and accounted for 74 % of the total. In particular, the increasing trend for processing and distribution corporations was pronounced, and of those, distribution corporations have greatly increased to 3,655 corporations in 2013, from 538 in 2000, 540 in 2005, and 1,730 in 2010 (Table 16.2).

3 The Policy Related to the Senary Industrialization of Agriculture in the ROK

It is possible to get a grasp, with the related policies for senary industrialization, of the policy projects being implemented with an objective of striving for the promotion of so-called agricultural-production-plus business, which fuses and combines secondary and tertiary industry upon the base of the primary industry of agriculture.

Regarding the configuration of senary industrialization, if there is the configuration of individually-run business diversification[3] (vertical and horizontal diversification), there will also be the configuration of business diversification via the linking-up of multiple entities. In addition, the configuration of senary industrialization will be diverse, including that a differentiation of types is possible depending on the central business.

In the ROK's agricultural policy, a variety of measures are being executed in concert with such a configuration of senary industrialization. Table 16.3 summarizes the related policy projects which have been carried out to date (Ministry for food, Agriculture, Forestry, and Fisheries 2012).

By the target for support (business entity), when there are instances with a target for support of individually-run entities, then there will be ones which presuppose

---

[2] In the 1990s, from the new perspective of rural amenities, attention was focused on rural tourism and urban–rural exchange to make practical use of the unique resources of rural areas for tourism. The requirement for such a rural tourism policy heightened with the 5-day working week system having been introduced, accompanying the increase in people's income in the late 1990s, and it was expected that the rural tourism market would grow in the future as well. In response to such developments, this has boosted, in policy terms also, the development of the rural tourism to date as a source of off-farm income.

[3] Yagi (2004) holds that in diversification, there are such things as: horizontal diversification which introduces production sectors other than vegetables and flowering plants in addition to paddy rice; vertical diversification which continues to expand the business sector in the direction of consumption including agricultural product processing and direct marketing; and diagonal diversification which expands into completely different business sectors based on existing production technology and sales channels, etc., like tourist farms and exchange business.

Table 16.2 Number of agricultural corporations by business category (2005–2013)

Category	Total agricultural corporations	Agricultural production	Other than agricultural production					
			Total	Processing	Distribution	Agricultural services	Farm management agencies	Other
2000	3,366 (100 %)	1,425 (42.3 %)	1,941 (58 %)	336 (10.0 %)	538 (16.0 %)	795 (23.6 %)	– (–)	272 (8.1 %)
2005	3,549 (100 %)	1,545 (43.5 %)	2,242 (63 %)	498 (14.0 %)	540 (15.2 %)	326 (9.2 %)	238 (6.7 %)	640 (18.0 %)
2010	8,361 (100 %)	3,112 (37.2 %)	5,618 (67 %)	1,568 (18.8 %)	1,730 (20.7 %)	613 (7.3 %)	369 (4.4 %)	1,338 (16.0 %)
2013	13,333 (100 %)	4,123 (30.9 %)	9,924 (74 %)	2,762 (20.7 %)	3,655 (27.4 %)	851 (6.4 %)	714 (5.4 %)	1,942 (1,406.0 %)

Source: *Agricultural and Fisheries corporation business entity statistics* for each year, Statistics Korea
Units: Business entities, %

Table 16.3 The policy related to the senary industrialization of agriculture

Support target (operating body)	Related policy project	Main content of support
Individually managed entities (individual farm agricultural corporations)	Farmers Small-Scale Start-Up Technology Support Projects	Funding support; consulting; etc.
	Rural Guest House Nurturing Projects	Funding support; consulting; etc.
	Direct Sales Point Facility Support	Funding support; consulting; etc.
	Educational Farm Nurturing Projects	Funding support; consulting; etc.
	Rural Restaurant Nurturing Projects	Funding support; consulting; etc.
	Rural Community Company Nurturing Projects	Funding support; consulting; etc.
Groups and organizations (comprised of plural entities)	Agricultural Product Comprehensive Processing Support Centers	Funding support; consulting; etc.
	Agriculture–Industry–Commerce Integrated SME Nurturing Projects	Funding support; promotion of links; consulting; etc.
	Invigoration Projects	Funding support; promotion of links; consulting; etc.
	Local Industry Nurturing Projects	Funding support; promotion of links; consulting; etc.
	Regional Strategic Food Industry Nurturing Projects	Funding support; promotion of links; consulting; etc.
	Rural Experiential *Maeul* Projects	Funding support; promotion of links; consulting; etc.
	Rural Theme Park Creation Projects	Funding support; promotion of links; consulting; etc.

Source: Compiled by the author from relevant materials

the configuration of project promotion groups, in which diverse local entities participate. For the former, support projects for small-scale start-up technology for farmers and nurturing projects for farm restaurants, etc., are appropriate, and primarily support for provision of funding and consulting is being undertaken. For the latter, invigoration projects and projects for nurturing regional strategic food industries, etc., are appropriate, and in addition to support for funding and consulting, etc., the promotion of links among the related entities is also included.

In this paper, from the above senary industrialization-related policy projects, I will raise "invigoration projects", which are business configurations in which multiple entities are linked, and "local industry cultivation projects", and summarize the current situation for the promotion of these projects and the challenges thereto.

3.1 Invigoration Projects

"Invigoration projects", within the course of modernization, industrialization, and urbanization, have the aim of seeking the invigoration of regional economies, by a focused supporting of rural areas lagging behind in development. The project targets, among the 234 fundamental local authorities nationwide, are the 70 cities and counties in the lowest 30 % in terms of population, industry, and public finances, and for a 3-year period approximately 10 billion won has been given in support to the selected cities and counties.

The special characteristic of the projects, differing from the top–down project promotion to date, is that the leadership authority for the project is granted to the local authority and they are allowed to conceive and promote freely the projects deemed necessary for the locality. Diverse bodies within the area, such as universities, firms, research institutes, NGOs, and media companies, become the project-promotion bodies, plan the development strategy in accordance with the peculiarities of the area, and support in terms of policy the carrying out of projects, such as the construction of the necessary networks for the strategy, production, and the creation of an income base, public services, welfare, and local marketing.

The content of the projects is in a configuration combining secondary and tertiary industry, on a basis of inter-sectoral convergence; in other words, agriculture and fisheries.

Taking a look at the 70 selected projects, a configuration for developing and selling processed products centered on local resources made up half, with 31 localities, followed by: tourism projects making use of local culture, with 12 localities; education and human resource development projects in accord with the peculiarities of the area, with 5 localities; medical projects making use of medicinal plants, etc., with 6 localities; development and cultivation projects for regional brands, with 10 localities; and projects making use of aquatic resources, with 5 localities (Table 16.4).

Table 16.4 The types of invigoration projects

Project type	No.	Projects selected
Development of local resources (specialty products)	32	Boseong county (green tea); Uiseong county (garlic); Naju city (pears); Goheung county (citrons); Damyang county (bamboo); Seongju county (*chamoe* Korean melons); Hoengseong county (Korean beef); Mungyeong city (*omija* berries); etc.
Regional cultural tourism	12	Yeongdong county (national music promotion); Gimje city (rice-farming culture); Inje county (leisure); etc.
Education and fostering of human resources	5	Geochang county (international education); Hapcheon county (Chinese classics education); Changnyeong county (foreign language education); etc.
Life and health industries	6	Jeongseon county (raw medicinal herbs); Jeungpyeong county (evening primrose); Geumsan county (ubiquitous health experiences); etc.
Regional image marketing	10	Pyeongchang county (Happy 700 Brand); Buyeo county (Goodtrae); Haenan county (Hwangto loess brand); etc.
Maritime aquatic resource development	5	Wando county (marine life industry); Sinan county (mudflat experiential tourism); Ulleung county (squid blue tours); etc.

Source: Compiled by the author referring to the "The situation for the promotion of invigoration projects" published by the Ministry of Government Administration and Home Affairs (2005)

However, although attempting to create projects, raising regional specialties and tourism and cultural resources, with just the local resources differing, there is no disparity in the content of the projects. In addition, while the key points in the construction of regional innovation systems are learning/education and networking, they have not progressed very much in the actual operation of projects, and for the most part it has been nothing more than one-way education from an invited speaker from those assembled or from outside. Moreover, it has had such problems as: complicated procedures overlapping with the existing regional development projects; the absence of bodies as leaders of innovation for regional projects; and the lack of the spontaneous participation of diverse regional bodies (Kim et al. 2005)

3.2 Local Industry Cultivation Projects

Local industry cultivation projects (from 2007) have an objective of striving for the invigoration of regional economies and the expansion of the income base, cultivating local resources with potential in rural areas toward compound industries with primary, secondary and tertiary industries linked together. To that end, they take project-implementing bodies as local communities composed of entities from industry, academia, government, and the research world, and place emphasis on project evolution via collaboration and cooperation among all the local bodies (Table 16.5).

For the projects a selection formula via public offering has been adopted, once selected as support targets they are brought in line with the project plans, and support is received to a scale of a total of 3 billion won over 3 years (50 % from national coffers, 50 % from the locality) as support for the software, such as the development of manufactured goods and brands, and marketing, and for the hardware, such as facilities and equipment. Every year, approximately 60 cities and counties apply for 70-odd projects, and from among them 30 are selected.

As for appraisal, evaluation groups composed of universities and research institutes, and private-sector experts, carry out written appraisals and site assessments. The main elements for evaluation are the situation for the utilization of local resources, the potential for industrialization, project promotion systems, and the validity of project plans, etc.

Moreover, targeted at the cities and counties which will become the promotional bodies for the selected projects, along with supporting consulting by experts in the project preparation period in order to encourage the promotion of more effective projects, and the undertaking, after the start of the projects, of monitoring of the project promotion situation and consulting by experts, they carry out annual evaluation of the project outcomes, and incentives and penalties are set.

Looking at the project groups which have been nurtured to date, in terms of project type (the results of 57 project groups among the project targets), they are, in order, food processing (65 %), non-food processing (14 %), experiential tourism (15 %), and others (6 %), with food processing accounting for more than half. Furthermore, within food processing there are general processed foods (33 %),

Table 16.5 The outcomes for local industry cultivation projects

Category	Annual turnover of participating firms (million won)			Employment generation (persons)			Increase in farmers' (residents') income (million won)			Export performance (US$ thousand)		
	2009	2010	2011	2009	2010	2011	2009	2010	2011	2009	2010	2011
2009 new businesses	247,853	260,410	271,645	1,521	1,798	2,264	77,140	158,470	145,427	33,945	46,870	26,746
2010 new businesses		108,717	129,049		1,063	962		47,961	66,800		294	1,922
2011 new businesses			67,806			423			97,055			15,889
Total	247,853	369,127	468,500	1,521	2,861	3,649	77,140	206,431	309,282	33,945	47,164	44,557
Average per business	8,852	6,592	5,512	54	51	43	2,755	3,686	3,639	1,212	842	524

Source: Materials in the Food Industry Policy Division, Ministry for Food, Agriculture, Forestry, and Fisheries

health and functional foods (20 %), kimchi (5 %), traditional foods and beverages (17 %), alcohol (15 %), and tea (11 %), and simple processed goods from regional agricultural produce predominate, and do not go far enough for the development of diverse business in which secondary and tertiary industries have combined. And, looking at the 3-year business performance of the new business group that was founded in 2009, it can be confirmed that improvement of the sales and the creation of employment of the participating companies, an increase in farm income (Table 16.5). Also, several entities are deployed overseas.

In addition, regarding project outcomes, from the survey conducted by the Regional Specialized Resources Commercialization Research Center (which targeted 47 project groups selected in 2007 and 2008) it was confirmed that certain outcomes were achieved, including: project groups in 42 localities achieving local employment generation; project groups in 22 localities achieving the attraction of related firms and start-ups; and project groups in 33 localities (85.6 % of the total) had turnover rise compared to the initial year. Moreover, it can be evaluated that in respect of the participation of local residents, there has been an increase from the initial year: the activity of study groups has increased from 50 instances in the first year, to 176 in the second year, to 224 in the third year; and the interchange and networking among local residents has been strengthening.

3.3 Challenges in Project Promotion

Next, the following two points will be raised as challenges in the promotion of these policy projects.

First, according to the results (Kim et al. 2011) from examining the experts' evaluations on policy projects by multiple entities (local communities), while projects for senary industrialization with a certain local ambit are striving for invigoration of local economies, by creating new business locally, based on mutual learning and cooperation through close networking among entities within the area, for ROK rural areas which are being confronted by the problems of the aging population and shortage of successors (young people in particular), with the actual situation of human resources to sustain the creation of such networking systems and opportunities lacking, there are many actual cluster project entities with configurations in formal fashion, and as for the promotion and management of projects in substance, it has been pointed out that there are many cases which are led by the authorities. Moreover, because the current situation is one where the linked relationships end up getting interrupted at the same time as the terminating of supported projects (Kim et al. 2005; Lee 2010), subsequently initiatives to maintain and strengthen the networking among entities are called for.

Second, without giving consideration to the feasibility of acquiring state subsidies, with a merit-based attraction of subsidies from state coffers in operation via a public-offering-style introduction, and with the problem of cases emerging which attract subsidies unreasonably or which determine their project plans via an external service aiming at their acquisition, the issue has been highlighted of deterioration ending up accelerating for those regions with a weak capability for making an application (Minki Jang et al. 2007).

4 Conclusion

In this paper, from the background to the promotion of the senary industrialization policy for agriculture in the Republic of Korea and the situation for the promotion of the related policy implemented to date, the things to examine regarding future challenges are as follows.

Within the senary industrialization policies which are being promoted for the sake of agricultural income and securing places of employment, there are instances which make a project target presupposition for project entities which are composed of multiple entities. This integrates and connects the production processes for agricultural produce and food items which becomes a division of labor, forms one kind of cluster, and attempts new value creation. However, nothing can arise in networks connected by mere lines. They must form new value (markets) derived from mutual interchange (collective learning), based on the sharing of a collaborative consciousness among each of the entities. Therefore, the related entities deepen the reciprocal exchange of information, and the forming of relationships of trust becomes an essential factor for the formation and endurance of clusters. In the future, it will be necessary for policy support to be devised relating to initiatives, etc., for making possible sustained linked relationships among the participating entities.

Additionally, what becomes the presupposition for the strengthening of networks among such entities is the strengthening of the capabilities of the participating entities, namely, the nurturing of human resources. In particular, a powerful leader with the strength to promote the project and the existence of cooperative partners to be its supporters are imperative.

Furthermore, what are called for are: planning and management of senary industrialization projects commensurate to the peculiarities and current local situation in each locality (hub) so as not to have regional skewing arising in the promotion of policy projects; a convergence of information; the establishment of an organization (platform) which is able to shoulder the role of intermediary, etc., between the interior and exterior of the network; and the cultivation of specialist human resources.

References

Futagami K, Hioki K (2008) The administration of cluster organization. Tokyo: Chuokeizai-sha, Inc. [in Japanese]
Jang Minki, Lim S, Kim K, Kang M, Kim Y, Lee Y, Chung S, Lee J (2007) Research into promotional schemes for the regional agricultural cluster vitalization project. Seoul: Regional Agriculture Network [in Korean]

Kim J, Park Y, Kim Y, Chung K, Kim K, Lim S (2005) Fact-finding survey into the promotion of regional agricultural cluster model projects. Seoul: Korea Rural Economic Institute [in Korean]

Kim J, Park S, Kim Y, Lim J (2011) Evaluation of and development schemes for the rural area vitalization policy. Seoul: Korea Rural Economic Institute [in Korean]

Lee Y (2010) The evolution of and future prospects for the regional agricultural cluster projects in the Republic of Korea. Agriculture No 1536, pp 70–75. Tokyo: Dainihon-Noukai [in Japanese]

Lee J (2012) Let's take a look at the base of agricultural issues. The Farmers Newspaper, 12 Jun 2012 [in Korean]

Lee Y (2013) The current state of and challenges for the senary industrialization of agriculture in the Republic of Korea: centered on the rural regional development policy. In: Overseas agricultural innovation policy and senary industrialization. Tokyo: Policy Research Institute. Ministry of Agriculture, Forestry and Fisheries (PRIMAFF) Project documents of supply chain, No 3, pp 97–138 [in Japanese]

Ministry for Food, Agriculture, Forestry, and Fisheries (2012) Agriculture and forestry business enforcement guidance document. Gwacheon: Ministry for Food, Agriculture, Forestry, and Fisheries [in Korean]

Ministry of Government Administration and Home Affairs (2005) The situation for the promotion of invigoration projects. Press material of policy. Regional Development Policy Bureau, Ministry of Government Administration and Home Affairs [in Korean]

Porter ME (1998) On competition. Boston: Harvard Business School Press [in English]

Rural Development Administration (2010) Fact-finding survey into the off-farm income activity accompanying the coming into force of legislation relating to support for the off-farm income activity of farmers. Suwon: Rural Development Administration and Jeong & Seo Consulting [in Korean]

Yagi H (2004) The modern Japanese agricultural business: management leading the way for the times. Tokyo: Association of Agriculture and Forestry Statistics [in Japanese]

Part III
Food Clustering in EU and North America

Chapter 17
Cluster Initiatives in Eastern Poland: Good Practices in Agriculture and Food-Processing Industry

Ewa Bojar, Matylda Bojar, and Wiktor Bojar

Abstract Although clustering processes are a relatively new phenomenon in Poland, regional cluster structures have already emerged in almost all regions. They are closely linked with the profile and economic specificity of regional economies. Regional clusters can be identified in almost all sectors of Poland's economy, including agriculture and food processing industry. There are indications that they are increasingly important for regional development and growth through making local and regional business establishments, including farms, more competitive and thus profitable.

Keywords Clustering processes • Agricultural clusters • Regional development • Agriculture and food-processing industry

1 Introductory Remarks

The concept of clusters is a new way of thinking about the economy and simulating its international competitiveness. This systemic approach assumes the non-linear nature of innovation processes. Clusters are a specific form of spatial organization of various manufacturing and service sectors. In terms of the ability to sustain economic development in the conditions of a post-industrial economy, they are considered the most mature form of organization of production (Szultka et al.2004, p. 81).

E. Bojar (✉)
Department of Economics and Management of Economy, Lublin University of Technology, Lublin, Poland
e-mail: e.bojar@pollub.pl

M. Bojar
Department of Management, Lublin University of Technology, Lublin, Poland

W. Bojar
Department of Small Ruminants Breeding and Agricultural Advisory, University of Life Sciences in Lublin, Lublin, Poland

Moreover, the clusters have the ability to stimulate development and sustain the competitive advantage of participating entities. There is an exceptional unanimity of opinions as regards potential benefits that the clusters can bring to local, regional and national economies. An effective cluster provides access to relatively inexpensive and specialized means of production and varied production resources and thus leads to the growth in productiveness of local businesses. Many researchers also argue that the spatial proximity of different companies and institutions stimulates innovative activities. Developing clusters are characterized by the growing number of start-ups, which translates into new jobs created in local economies (Szultka et al.2004, p. 81).

Observations of existing clusters justify the claim that they can be an effective driver of regional development, creating conditions conducive to the dissemination of innovative technologies and thus increasing the international competitive advantage of regional firms. Szymoniuk (2003) noted that cooperation between farmers, enterprises and scientific institutions in clusters can produce additional benefits for participating entities in the form of increased profits, bigger market share, implementation of innovative production methods, as well as product and management innovations. Many authors also argue that regional cluster structures attract foreign investments and contribute significantly to the increase of export sales (Bojar et al. 2007, p. 40). Cooperation within the framework of clusters produces synergy effects for the cluster participants and thus allows them to achieve competitive advantage in highly demanding markets. Moreover, Bojar and Stachowicz (2008) observed that small and medium-sized enterprises often form clusters in order to apply jointly for external funds available under various EU programmes.

Many researchers have long been trying to find the distinctive features of cluster structures, which has resulted in the current abundance of cluster definitions available in literature. Although different aspects of the clusters are brought to the fore, most of available definitions specify the following elements which constitute the cluster (Grycuk 2003, p. 4):

- Geographical (spatial) concentration. This spatial proximity is conducive to the spread of innovations. It also encourages development of relations and cooperation between partners.
- Cooperation and competition. Very often cooperation and competition between cluster members are analysed jointly and inseparably as these phenomena reinforce the innovation potential of cluster participants and produce synergy effects.
- Sectoral concentration – cluster participants often operate in the same or related sector(s).
- Specialization, which is conducive to raising the efficiency of organizations and stimulates their organizational improvement. Simultaneously, specialization reinforces the need for cooperation and stimulates the development of networks of cooperative links between companies.
- Interdependence. Various dynamic interactions take place between cluster participants, and the quality and intensity of these interactions are the key determinants of the cluster's economic success.

Moreover, some researchers put strong emphasis on what they call the 'joint development trajectory' (co-evolution) of the cluster, operating on the same markets, technology or common knowledge base, which very frequently is a public good. Some authors also note that equally important is a sense of regional community and a common vision for the cluster's development, accepted by all cluster participants (Grycuk 2003).

Although clustering processes are a relatively new phenomenon in Poland, the cluster structures have already emerged in almost all regions. As noted by Bojar (2009), they are closely linked with the profile and economic specificity of the region in which they appear.

2 A Brief Characteristics of Poland's Agricultural Potential

The total area of Poland is 31,268,500 ha, of which 61.2 % is arable land. About 38.6 % of the country's population live in rural areas, which account for 93.2 % of the total area of Poland. Most farms are relatively small businesses: 72.5 % of farms are smaller than 5 ha. They cultivate approx. 18.3 % of total arable land. Farms with more than 20 ha of arable land represent only 4.3 % of farms; however they cultivate 43.2 % of all arable land in Poland. The remaining 40 % of farms are medium-sized, with 5–20 ha of land.

Farming is naturally linked to the food-processing industry which is the main and direct purchaser of agricultural products. The financial standing of food-processing companies has a direct impact on the farmers' revenue. The scale effect can only be achieved when these sectors are properly managed in a coordinated manner. Currently, the existing links between the primary production and industry (including wholesale) should be strengthened in order to increase the competitiveness of Polish agriculture (KROW 2010, p. 60). The Rural Areas Development Programme, covering 2007–2013, assumes that extensive actions shall be undertaken in order to encourage and stimulate development of cooperation between farmers and food-processing businesses and wholesalers, preferably in the form of various groups of agricultural producers (PROW 2010, p. 298). These groups tend to focus on improving the farmers' position on the market, their financial standing (e.g., through the reduction of costs), improving the quality of delivered produce (e.g., through shared technologies), joint planning, shared logistics and concentration of deliveries, and finally more flexible response to market needs and requirements, and therefore they constitute a natural base on which agricultural clusters could develop and build on. Hence the groups of agricultural producers have a crucial role to play in the process of cluster formation and development.

3 A Brief Characteristics of the Agriculture and Food-Processing Sector in Eastern Poland

In 2007, the Ministry of Regional Development commissioned research on the clusters and clustering processes in eastern Poland. The term 'eastern Poland' refers to the sub-region which encompasses 5 large administrative units (voivodeships), divided into 101 smaller administrative units (poviats). Lublin Voivodeship includes 20 poviats and 4 cities (urban poviats), Podkarpackie Voivodeship includes 21 poviats and 4 cities, Podlaskie Voivodeship – 14 poviats and 3 cities, Swietokrzyskie Voivodeship – 13 poviats and 1 city, and Warmińsko-Mazurskie Voivodeship includes 19 poviats and 2 cities. Eastern Poland, with its agriculture being a dominating sector of regional economy, is one of the most underdeveloped and economically backward regions in Poland. Statistical analysis of economic indexes in all poviats of the region, carried out within the framework of the research on the development of cluster structures in Eastern Poland, clearly shows that the food-processing industry, which employs approx. 30 % of regional population, is a dominant sector of eastern Poland's economy (Plawgo 2007b), cf. Fig. 17.1).

Regional concentration of industries indirectly indicates that there are at least several so-called 'cluster-like concentrations', which according to Porter's theory can be considered 'cluster germs' (Porter 2001, p. 246). Therefore, a more precise identification of these 'cluster germs,' that may or may not develop into clusters, is crucial for any cluster-oriented policy to be successful (Plawgo 2007b, p. 9). As Plawgo (2007b) noted in his report on the development of cluster structures in eastern Poland '(…) A cluster germ can be defined as a concentration of various entities, in particular enterprises involved in manufacturing and delivering certain products and/or services, that have already established some cooperative links between them. Once those entities demonstrate strong willingness to develop their widely understood cooperative relations in order to improve the competitiveness of the whole group, the so-called 'cluster initiative' can be implemented. (…) A cluster initiative can be defined as intentional efforts undertaken in order to improve the competitiveness of the group and involved firms through constantly developed cooperation between all partners, that is businesses, government and self-government bodies, universities, research-and-development units, etc. Cluster initiatives are undertaken either in order to expedite the process of cluster formation or strengthen already existing clusters. (…) Cluster initiatives are something quite different than objective social and economic structures like clusters. (…)'. Cluster initiatives emerge mainly on the areas with a high concentration of companies operating in similar or related sectors (Table 17.1).

4 An Overview of the Main Agricultural and Food Clusters Identified in the Regions of Eastern Poland

4.1 Agricultural and Food Clusters in Podlaskie Voivodeship

Podlaskie Voivodeship is located in the north-eastern part of Poland. Its overall area is 20,187 sq. m. which accounts for about 6.5 % of the total Poland's area. The regional economy is based on the food-processing industry, light industries, wood-working, construction, as well as on engineering and the machine-building industry.

17 Cluster Initiatives in Eastern Poland: Good Practices in Agriculture...

Fig. 17.1 The concentration of agriculture and the food-processing industry in poviats of Eastern Poland. (**a**) Production of foodstuffs and beverages, (**b**) production of foodstuffs and beverages, farming, hunting, and related services (Source: A report on research on the development of cluster structures in eastern Poland)

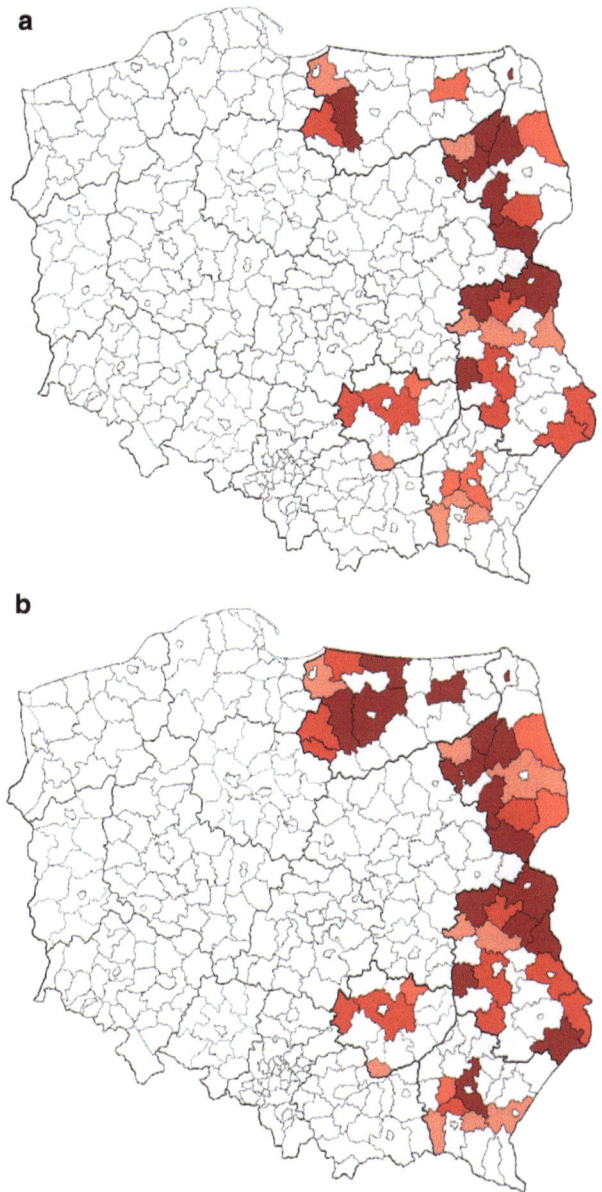

In recent years the food-processing sector, processing local produce, has been growing rapidly. Regional agriculture is based on raising livestock; dairy and meat products account for almost 90 % of the total regional production output (Plawgo 2007a, p. 45). In Podlaskie Voivodship there are several breweries and big companies processing locally produced milk, meat, and poultry.

Table 17.1 Cluster initiatives and cluster germs identified in agriculture and the food-processing industry in Eastern Poland

	Cluster's name	Location
1	Klaster Mleczarski [Dairy cluster]	Warmińsko-Mazurskie Voivodeship
2	Klaster Browarów Regionalnych [The cluster of regional breweries]	Warmińsko-Mazurskie Voivodeship
3	Klaster Mięsa Wołowego [Beef cluster]	Warmińsko-Mazurskie Voivodeship
4	Podlaski Klaster Spożywczy [Food cluster of Podlasie]	Podlaskie Voivodeship
5	Klaster Spożywczy "Naturalnie z Podlasia" [Foodstuffs cluster "produced in Podlasie"]	Podlaskie Voivodeship
6	Podlaski Klaster Piekarniczy – Stowarzyszenie [Bakery cluster of Podlasie – Association]	Podlaskie Voivodeship
7	Klaster Mleczny "Podlaskie" [Dairy cluster "Podlaskie"]	Podlaskie Voivodeship
8	Dolina Ekologicznej "Żywności" [Ecological food valley]	Lubelskie Voivodeship
9	Chmielaki Nadwiślańskie "Lubelskie" [Lublin cluster of Hops Growers]	Lubelskie Voivodeship
10	Klaster Spożywczy [Food cluster]	Lubelskie Voivodeship
11	Podkarpackie Smaki [Podkarpackie flavours]	Podkarpackie Voivodeship
12	Klaster Winny [Wine cluster]	Podkarpackie Voivodeship
13	Klaster Ogrodniczo-Sadowniczy [Horticulture and orcharding cluster]	Świętokrzyskie Voivodeship
14	Pomidor z Ziemi Sandomierskiej [Tomatoes from the Sandomierz land]	Świętokrzyskie

Source: Plawgo (2007b), p. 325

4.1.1 Case Study: The Cluster Initiative 'Food Cluster of Podlasie'

One of the most important cluster initiatives which emerged in Podlaskie Voivodship is 'Food Cluster of Podlasie'. This innovative research-and-development initiative operates in the key sector of the regional economy – food processing – and has received significant financial support within the framework of The Integrated Operational Programme of Regional Development, Action 2.6. 'Regional Innovation Strategies and Knowledge Transfer' (Plawgo and Klimczuk 2007). Activities of the cluster are focussed on improving the innovativeness of the regional food and food-processing industry in terms of used technologies, offered products and organizational solutions. The cluster also has an ambitious strategic goal – setting up a regional food-processing research-and-development centre. The cluster is actively engaged in promoting regional and local foodstuffs on a national level as a sort of 'jewel in the crown', and provides a wide spectrum of training courses. Participants of this initiative have easier access to innovative solutions, which boost their capacity to acquire new markets, both domestic and overseas (Plawgo and Klimczuk 2007).

One of the first activities undertaken by this innovative cluster initiative was the process of cluster mapping which enabled identification of major geographical concentrations of companies operating in food-related sectors. Based on the cluster mapping, four regional sub-clusters have been identified, i.e., a meat cluster, fruit and vegetable cluster, cereal and bakery cluster, and dairy cluster. Then companies which potentially could be participants of identified sub-clusters were included in benchmarking studies in order to assess their innovativeness (Plawgo and Klimczuk 2007).

Efforts undertaken within the framework of this initiative increased people's awareness of the need to cooperate and develop lasting cooperative links based on trust and mutual confidence. In 2008, when financial support under the EU programme ended, the cluster ceased operations. However, at the end of 2013, some companies which formerly participated in the cluster joined another cluster initiative – Industrial and Food-processing District of Podlasie, which assembles food companies from three voivodeships of Eastern Poland (GB 2013).

Foodstuffs Cluster 'Produced in Podlasie' is a prime example of a bottom-up initiative of local companies which decided to establish and develop formal cooperative links in order to achieve synergy and scale effects. As distinct from the initiative Food Cluster of Podlasie, the process of formation of the Foodstuffs Cluster 'Produced in Podlasie' was initiated by one of the leading firms operating in the regional food sector, Agrovita, which had already cooperated with other participants in this cluster initiative (e.g., product cooperation).

The Foodstuffs Cluster 'Produced in Podlasie' was formally registered as an association of natural persons. The association has 12 founder members representing 9 companies operating in the food sector, 1 trading company (distributor), 1 packaging vendor, and 1 IT company. Moreover, the association has 2 supporting members – The Chamber of Industry and Trade in Łomża and The Marshal's Office of Podlaskie Voivodeship.

Products offered by the cluster members were sold under a common brand. The cluster has also developed a joint cluster development strategy and export strategy based on the enormous financial potential of companies participating in the initiative, a huge production capacity, mutual confidence between participating companies and their awareness of the need to develop strong and lasting cooperative relations in order to pursue common goals. Unfortunately, due to the lack of financial support, the cluster gradually ceased activities (Plawgo and Klimczuk 2007).

4.2 Clusters Operating in Świętokrzyskie Voivodeship

Świętokrzyskie Voivodeship is located in central-eastern Poland. It is one of the country's smallest voivodeships with a total area of 11,708 sq. kilometres, accounting for 3.7 % of the overall area of Poland. Świętokrzyskie Voivodeship has approx.

1.3 million inhabitants, which represent about 3.3 % of Poland's total population. The regional economy is based on agriculture and industrial production. Agriculture is focused in the southern parts of the region, and the basic crops are cereals and potatoes.

4.2.1 Case Study: The Cluster Initiative 'Tomatoes from the Sandomierz Land'

The beginning of this initiative dates back to 2005 when a training project promoting clustering was started. The project was commissioned by the Polish Agency for Enterprise Development (PAED) and was subsidized by the European Union within the framework of the European Social Fund. The cluster initiative *Tomatoes from the Sandomierz Land* was introduced in the area of Sandomierz, a small city in Świętokrzyskie Voivodeship, and was based on a regional product, a special species of tomatoes grown around the city and took advantage of local climate and soils conditions (Olesiński and Predygier 2007).

The partners to this initiative were selected vegetable farms, seed vendors, vendors of fertilizers and other chemicals used in tomato growing (herbicides, pesticides and fungicides), scientific institutes, certification units, tomato purchasing and packaging centres, agricultural machinery vendors, consulting firms, marketing and PR agencies, as well as local self-government authorities concerned with the development of local tomato growers and food-processing firms operating within the tomato cluster. The cluster sold fresh tomatoes to domestic and international customers (Olesiński and Predygier 2007).

Activities carried out within the framework of the project have raised social awareness of the importance and benefits resulting from the cooperation within the cluster, and enabled knowledge transfer and development of up-to-date farming technologies. This cluster initiative failed to develop stable and lasting cooperation links and the cluster ceased operations shortly after the EU subsidies has ended.

4.3 Agricultural and Food Clusters in Lublin Voivodeship

In terms of the occupied area, Lublin Voivodeship with the total area of over 25,000 sq. kilometres is one of the largest voivodships in Poland. It represents 8 % of the total Poland's area. The region has approximately 2.2 million inhabitants and the population density is equal to 71.3 % of the national average. The urbanization index equals 46.7 % of the country's average (Bojar 2006). According to official data published by the Central Statistical Office (2012), gross domestic product per capita generated in Lublin Voivodeship in 2010 was 25,079 Pln (calculated in current prices), which represented only 67.6 % of Poland's GDP per capita and only 25 % of GDP per capita recorded in the European Union.

These data alone clearly show that Lublin Voivodeship with its ailing and largely labour-intensive economy is one of the most underdeveloped regions in Poland and the European Union. The regional economy is based on agriculture and outdated industry with relatively low investment outlays. Regional economic structure is underdeveloped and underinvested. Moreover, the region has the most scattered urban network in Poland (Bojar 2007, p. 152). Production represents only 10 % of all economic activities carried out in the region, and the regional industrial output accounts for as little as 2.9 % of the total industrial production in Poland (Krasowicz 2004).

Regional specialization of economy is largely based on science and tertiary education, ecological farming, food-processing, tourism, and IT technologies (Plawgo 2007b). Agriculture is the most developed sector of regional economy. Its share in Poland's overall agricultural production is 8.3 %, which is 3.5 times higher than the share of the regional economy in industrial production. In terms of the total acreage used for agricultural production, the region is ranked third in the country.

The region is a significant producer of various species of cereals, potatoes, sugar beets, vegetables, hop, tobacco and hemp. Regional agriculture is a major producer of milk, honey, a wide variety of herb species, as well as soft fruits, especially raspberries. Moreover, animal breeding is also an important part of regional agriculture, in particular swine, cattle and horse breeding. Food-processing is one of the most resilient and prosperous sectors of the regional economy. Its sales accounts for almost 30 % of the total regional sales volume of products and services. It is beyond question that the quality of offered products is behind the success of this regional food-processing sector.

Paradoxically, poor industrialization of the regional economy is one of the most valuable assets of the Lublin region. Low levels of industrial pollutants, traditional forms of farming, favourable climatic and environmental conditions, in particular a huge percentage of extremely fertile soils, as well as favourable terrain configuration are those assets that the regional economy can build on in order to become the leading producer of ecological foodstuffs.

Constantly growing demand for organic food observed in recent years opens new perspectives for agriculture in the region. However, organic farms account for as little as 0.11 % of all farms in Poland. In the years 1999–2004, the overall number of certified ecological farms in the Lublin region has nearly doubled from 263 to 466. Eight out of the 55 food-processing plants operating in Poland which were certified to be organic in 2004 are based in the Lublin region (Szymoniuk 2007).

4.3.1 Cluster Initiative Case Study: Ecological Food Valley/Association EkoLubelszczyzna – The Cluster of Clusters

The cluster 'Ecological Food Valley' is one of the most significant regional cluster initiatives in Lublin Voivodeship. The foundations for the cluster formation have been laid by the EU-financed project 'Strategy for the Ecological Food Valley'.

The project was completed in 2006, and in January 2007 the Association 'EkoLubelszczyzna' was formally established. The idea of the formation of an 'Ecological Food Valley' in the Lublin region and neighbouring voivodships was born at the ministerial level in the first years of the twenty-first century. The project has been incorporated into the National Development Strategy and became part of the Lublin Voivodeship Development Strategy as well (Bojar and Bis 2007). The very idea of formation of a regional ecological food cluster was inspired by the fact that in developed countries ecological farms deliver at least 10 % of the total food supply and, in the case of some foodstuffs items, a share of organic food reaches even 30 %, while in Poland the sales of organic food accounts for as little as 1 % of the total food sales volume. In Poland, as of the end of 2007, 11,887 farms used organic methods of production, which is a 29 % growth as compared to 2006. 5,159 of these farms were just in the midst of the process of switching to fully ecological methods of food production. In the same period we also saw a 27 % increase in the number of green food producers. A peak 81 % annual increase in the number of organic food-processing plants was recorded in 2006 (Wieczorkiewicz 2007). At the end of 2007 there were 207 certified food-processing plants in Poland processing produce from organic farms. In 2012, the number of certified ecological food-processing plants in Poland increased up to 312, compared to 270 plants operating in 2011. As of December 31, 2012, in Poland operated nearly 26,000 organic farms. As compared to 2011, both the total area used by ecological farms and the number of ecological farms increased by approximately 10 % (Przybylak 2013).

These figures clearly show that in Poland both the potential and capacities for ecological food production still remain unexploited. Activities performed by the cluster build on the regional strengths and are in line with the adopted strategy and the provisions of the Regional Innovation Strategy and the National Development Strategy 2007–2015.

The association 'EkoLubelszczyzna', which was intended to be 'the cluster of clusters', undertakes various initiatives aimed at stimulating both supply and demand for ecological food and increasing the ecological awareness of society at large. It is also actively engaged in the promotion of the region.

The association delivers a wide range of training sessions for its members and provides them with specialized advice and consulting on legal, economical, organizational, and technological issues. 'EkoLubelszczyzna' organizes conferences and various contests promoting ecological approach and healthy lifestyles. In order to encourage cooperation between partners and the establishment of new business contacts, the association has developed and maintains a dedicated information exchange system. Moreover, 'EkoLubelszczyzna' actively participates in social consultations and acts as a think-tank for various administrative and governing bodies (Bojar and Bojar 2007).

4.3.2 The Cluster Initiative 'Lamb from the Lublin Region'

The present shape and the structure of this cluster initiative grouping regional lamb breeders has been formed by a group of nearly 130 farmers, the owners of small animal farms and members of the Regional Union of Sheep and Goats Breeders (Bojar and Bojar 2013). The three main goals of the breeders are: the welfare of grown animals, the quality of meat delivered to the market, and the maximization of meat production. The Union also groups farmers involved in conservation breeding.

The union is actively involved in promoting good breeding practices and innovative breeding technologies. Its members participate in various regional and nationwide events, including trade shows, contests and exhibitions. The union cooperates with many meat-processing plants and exporters.

The cluster initiative 'Lamb from the Lublin Region' and the strategy for the development of the regional animal production sector build on the clustering potential of the Regional Union of Sheep and Goats Breeders. This cluster initiative has laid the foundations for the creation of the frameworks for broader and more productive cooperation between lamb breeders and regional research-and-development units, scientific institutions, production certification units, agricultural associations, and regional authorities, which have contributed to closer multi-stage integration facilitating faster and more efficient implementation of innovation, effective exchange of information and knowledge diffusion, as well as the dissemination of good animal breeding and meat production practises. All these activities create better prospects and business development opportunities for the regional lamb breeders and meat producers and contribute significantly to increasing their competitive advantage in both domestic and international markets (Bojar and Bojar 2013).

5 Summary

The cluster is an objective phenomenon resulting from the geographical concentration of various entities operating in the same or related sectors, which bestow trust upon each other, and cooperate and compete at the same time. The awareness of objectively existing links between these entities is a good starting point for joint actions to develop their competitive advantage in the market. The clusters identified in Eastern Poland are in a different stage of development. The current trends in European economic policy and corresponding financial support have triggered many more or less successful cluster initiatives. In addition to this support, many regional governments have strategies providing financial support to regional clusters operating in the key sectors of regional economies. However, as Plawgo (2007a) argues, actions undertaken within the framework of adopted economic development policies can only reduce development barriers or strengthen existing clusters by providing them with needed services or other resources, and cannot substitute market mechanisms which create and shape the objective conditions for the emergence and development of the given cluster in the given area.

Eastern Poland is one of the poorest macro-regions of the European Union. The clusters and potential clusters (so-called cluster germs) presented in this paper directly take advantage of spatial concentration in the key sectors of the regional economy. In all analysed cluster initiatives we observed the process of constant development of cooperative links between the primary businesses operating in a given sector and other firms offering needed services, as well as regional self-government and local authorities. Despite the fact that these clusters operate in the same industry, i.e., food-processing, they differ in many ways, in particular in terms of the setting-up process, legal forms, performed operations and territorial range.

In eastern Poland we have identified the following three types of clusters:

classical production clusters ('Tomatoes from the Sandomierz Land' in Świętokrzyskie Voivodeship and foodstuffs cluster 'Produced in Podlasie');
clusters which group their members around common ideas, for instance the promotion of certain production methods and lifestyles ('Ecological Food Valley' in Lublin Voivodship); and
clusters that play a role of specific advisory centres ('Food Cluster of Podlasie', 'Lamb from the Lublin Region').

Some clusters have emerged as a result of a bottom-up initiative (e.g., foodstuffs cluster 'Produced in Podlasie'), while the establishment of the other clusters results directly from previously developed regional programme documents, for example the cluster 'Ecological Food Valley' was established based on the provisions of adopted Regional Development Strategy.

Even though most of these clusters have been operative for more than 5 years now, it is very difficult to assess tangible and measurable economic effects for the regional economy that could be attributed directly to these clusters. However, what is beyond question is the fact that at least one natural barrier has already been overcome. Members of these clusters, previous competitors, have learnt that collaboration can bring even more benefits to them than competition alone and that cooperation is a factor upon which they should build their competitive advantage. Although some of these clusters ceased operations due to the lack of financial support, one can observe new cluster initiatives emerging in the key sectors of regional economy.

References

Bojar M (2009) Klastry w polityce regionalnej – przykład wojewodztwa lubelskiego. In: Kowalczewski W, Matwiejczyk W (eds) Aktualne problemy zarządzania organizacjami. Difin, Warsaw

Bojar E (ed) (2006) Klastry jako narzędzia lokalnego i regionalnego rozwoju gospodarczego. Lublin University of Technology Press, Lublin

Bojar W (2007) The role of groups of agricultural producers in the process of cluster-formation in the Lublin region. In: Bojar E, Olesiński Z (eds) The emergence and development of clusters in Poland. Difin, Warsaw

Bojar E, Bojar M (2007) Wyniki badań w wojewodztwie lubelskim. In: Plawgo B (ed) Badanie struktur klastrowych Polski Wschodniej. Ministry of Regional Development, Warsaw. http://www.funduszestrukturalne.gov.pl/NR/rdonlyres/86F295DC-2C0A-4A1C-BAD2-5E73122D9C3D/45027/RozwjstrukturklastrowychwPolsceWschodniejRAPORT.pdf. Viewed 17 Dec 2013

Bojar E, Bis J (2007) Major threats to economic clusters in Poland. In: Bojar E, Olesiński Z (eds) The emergence and development of clusters in Poland. Difin, Warsaw

Bojar E, Stachowicz J (2008) Konstruowanie dynamiki procesow poznawczych w organizacjach iregionie racjonalizacją organizowania rozwoju sieci. In: Stachowicz J, Bojar E (eds) Sieci proinnowacyjne w zarządzaniu regionem wiedzy. Lublin University of Technology Press, Lublin

Bojar W, Bojar E (2013) Development strategy of the cluster the lamb from the Lublin region. A report from research carried out within the framework of the project financed by the National Centre for Research and Development "The one-year-cycle lamb meat production for the domestic market needs"

Central Statistical Office (2012) Rocznik Statystyczny Wojewodztw/Statistical yearbook of the regions – Poland [billingual publication]. CSO, Warsaw. ISSN 1230–5820. http://www.stat.gov.pl/cps/rde/xbcr/gus/rs_rocznik_stat_wojew_2012.pdf. Viewed 17 Dec 2013

Gazeta.pl Białystok GB (2013) http://bialystok.gazeta.pl/bialystok/1,35241,13758266,Nowy_klaster_Polski_Wschodniej__Podlaski_Okreg_Spozywczo_Przemyslowy.html#ixzz2lpEB8G7g. Viewed 4 Dec 2013

Grycuk A (2003) Koncepcja gron w teorii i praktyce zarządzania, Organizacja i Kierowanie nr 3

Krasowicz S (2004) Czynniki ograniczające wykorzystanie potencjału rolnictwa Lubelszczyzny. Biuletyn Informacyjny nr 9

KROW (2010) Kierunki rozwoju obszarów wiejskich założenia do „strategii zrównoważonego rozwoju wsi i rolnictwa. Warszawa. http://www.lubelskie.pl/img/userfiles/files/PDF/Rolnictwo/KROW.pdf

PROW (2010) Program rozwoju obszarów wiejskich 2007–2013. Ministry of Agriculture and Rural Development Warsaw. http://www.prow.sbrr.pl/pliki/PROW_2007-2013_pazdziernik_2010.pdf. Viewed 15 Dec 2013

Olesiński Z, Predygier A (2007) Wyniki badań w wojewodztwie świętokrzyskim. In: Plawgo B (ed) Badanie struktur klastrowych Polski Wschodniej. Ministry of Regional Development, Warsaw. http://www.funduszestrukturalne.gov.pl/NR/rdonlyres/86F295DC-2C0A-4A1C-BAD2-5E73122D9C3D/45027/RozwjstrukturklastrowychwPolsceWschodniejRAPORT.pdf. Viewed 15 Dec 2013

Plawgo B (2007a) Potencjał rozwoju regionalnego – wojewodztwo podlaskie. Białostocka Fundacja Kształcenia Kadr, Białystok

Plawgo B (ed) (2007b) Rozwoj struktur klastrowych Polski Wschodniej. Ministry of Regional Development, Warsaw. http://www.funduszestrukturalne.gov.pl/NR/rdonlyres/86F295DC-2C0A-4A1C-BAD2-5E73122D9C3D/45027/Rozwjstrukturk lastrowychwPolsceWschodniejRAPORT.pdf. Viewed 15 Sept 2012

Plawgo B, Klimczuk M (2007) Wyniki badań w wojewodztwie podlaskim. In: Plawgo B (ed) Badanie struktur klastrowych Polski Wschodniej. Ministry of Regional Development, Warsaw. http://www.funduszestrukturalne.gov.pl/NR/rdonlyres/86F295DC-2C0A-4A1CBAD2-5E73122D9C3D/45027/RozwjstrukturklastrowychwPolsceWschodniejRAPORT.pdf. Viewed 15 Dec 2013

Porter ME (2001) Grona a konkurencja. In: Porter ME (ed) Porter o konkurencji. PWE, Warsaw

Przybylak K (2013) Polski rynek żywności ekologicznej w liczbach, Biokurier. http://biokurier.pl/aktualnosci/2287-polski-rynek-zywnosci-ekologicznej-wliczbach. Viewed 17 Feb 2014

Szultka S, Brodzicki T, Wojnicka E (2004) Klastry: Innowacyjne wyzwanie dla Polski. Instytut Badań nad Gospodarką Rynkową, Gdańsk

Szymoniuk B (2003) Klastry wiejskie na Lubelszczyźnie – praktyka grupowej przedsiębiorczości. Organizacja i Kierowanie Nr 2(112), p 114

Szymoniuk B (2007) L'agriculture francaise et l'agriculture polonaise dans l'Europe de 2007: experiences partagees et interets communs. http://www.ekolubelszczyzna.pl/artykuly.php?art=0&queryId=3. Viewed 30 Sept 2012

Wieczorkiewicz R (2007) Rynek żywności ekologicznej w Polsce. Polski Portal Finansowy Bankier.pl. http://www.bankier.pl/wiadomosc/Rynek-zywnosciekologicznej-w-Polsce-1738709.html. Viewed 20 Mar 2007

Chapter 18
Main Factors Affecting Food Industry Clustering in France

Nejla Ben Arfa and Karine Daniel

Abstract This paper provides insight into the spatial structure of the French food sector by investigating the forces of agglomeration and dispersion that influence the location of food industries in France in 2007. We use Spatially Weighted two Stage Least Squares estimation, suggested by Kelejian and Prucha (2007), which enables us to analyze both types of endogeneity (spatial lag and Cluster).

Results show that no agglomeration effect exists beyond the employment zone, and that a competition effect about attractiveness between those geographical areas rather exists. This is inherent to the definition of employment zones in France which are determined by the economic activity that develops within them.

Results also show that agricultural employment, infrastructure and urbanisation are the main factors explaining the clustering of the agri-food industries. It is interesting to note that the influence of clusters policy increases this agglomeration.

Keywords Food sector • Spatial econometrics • Agglomeration • Clusters policy • France

1 Introduction

The Food Industry is an important economic sector in France. Indeed, France is one of principle food producers and exporters in the world. Food industry is at the forefront of France's industrial sector in terms of employment (575,900 employees in 2011), annual turnover (169 Billion euros in 2011) and added value (36 Billion euros). In 2011 it made 17 % of the added value for the whole manufacturing sector.

Food industry lies between, and is therefore strongly linked to, the agricultural and distribution industries, both major components of the economy. France, Europe's largest producer of agriculture, is well positioned to supply the agroalimentary sector with raw materials. Production is spread across the land with

---

N. Ben Arfa • K. Daniel (✉)
Social Sciences Laboratory, Ecole Supérieure d'Agriculture ESA, Angers, France
e-mail: n.benarfa@groupe-esa.com; k.daniel@groupe-esa.com

© Springer Japan 2016
L. Kiminami, T. Nakamura (eds.), *Food Security and Industrial Clustering in Northeast Asia*, New Frontiers in Regional Science:
Asian Perspectives 6, DOI 10.1007/978-4-431-55282-6_18

different regions specialising in different areas of production. This widespread production lends to explain the location of different agribusinesses, along with economic and demographic factors (Home Market Effect) such as income and unemployment (Bagoulla et al. 2010).

Within this sector, geographical spread is particularly pronounced. Brittany and Pays de la Loire (West France) alone account for 25 % of the nation's agribusiness workforce. This lends to a clustering effect. Regroupment of French agribusiness at a local level is evident. Spatial concentration of economic activities in dynamic territories is encouraged by the government. The idea being that agglomeration encourages competitivity. The implementation of competitive clusters is aimed at reinforcing the agglomeration of industrial activities.

The *cluster* concept is increasingly used to enhance the economic momentum of territories that compete with one another. The work of Porter (1998, 2000) has been very influential in this matter and has been used as justification for cluster policies. A typical defense of cluster policies is that clusters bring economic gain and should, therefore, receive public support. France has developed a policy based on this concept, which became a pan-European trend following the Lisbon strategy defined in 2000 and laid out in the 2020 European plan. This strategy seeks to make the European Union a competitive economy through the development of knowledge. Under the strategy, governments are encouraged to increase expenditure devoted to research and development for innovation (objective of 3 % of GDP). Hence, France is adopting a specific policy, based on the cluster models set out in the literature, to enhance the territories' economic development through the establishment of competitive clusters [*pôles de compétitivité*]. In addition to this national-level effort, local authorities, and especially 22 specific regions with their own economic development responsibilities, are also investing in the development of such clusters.

The French Government's *competitive clusters policy* was adopted in 2005. The first phase was a call for proposals to give them official accreditation. In 2005 the selection lists 66 competitive clusters, which rose to 71 in 2009. Six of these Competitive clusters are directly dedicated to Agri-food activities.

Studying the French case is interesting because there is a long tradition of strong government intervention regarding the location of economic activities. This is particularly evident in agriculture because French cluster initiatives are more or less unified across the country (Duranton et al. 2010). Until the implementation of competitive clusters, French spatial planning policies aimed at avoiding the concentration of economic activities over a few territories and guaranteeing equity between territories. Now, collaboration and spatial concentration of economic activities in dynamic territories are encouraged by the government. The aim is to achieve a certain "critical mass" in order to remain internationally competitive and to build collaborative projects enabling companies to innovate and position themselves at the forefront of their sectors, both in France and abroad. Equity remains a consideration, as these cluster policies are well distributed across the country and represent most economic sectors (Ben Arfa et al. 2013).

Cluster policies in France have become very popular. However, the determinants of implemented cluster policies are not clear and the supported collaborations are not examined. As Duranton et al. (2010) state *"cluster policy (..) requires to pick not* only the 'right' industries but also the 'right' territories. It is interesting to note that one of the fathers of Silicon Valley, Frederick Terman, who was the vice president of Stanford University, was unable to replicate this experiment in New Jersey a few years later when called upon by the Bell Laboratories (Leslie and Kargon, 1996). There exist actually very few examples of public policies that were successful in promoting clusters"* (Duranton et al. 2010, p. 4). In this respect, the question regarding the French competitive cluster policy is: is there any evidence, within the considered sector, that the selected location of the cluster actually grouped more agglomerated activities?

To analyze how food industries are localized in France, and to identify the main factors of the localisation of this sector, especially to see if cluster policies are concerned with the right sector and industry and implemented in the right territory, we used explanatory spatial data analysis and spatial econometric regression to detect the spatial structure and dynamics of agri-food industries and connect them to the location of the competitive clusters. The objective of this paper is to determine the main factor of localization of food industries and to identify the impact of cluster policies implemented by the French government.

In the first part, we present the spatial dynamics and location of the food industries in France and the official food industry clusters supported by the French government. In the second part we propose to determine the main factors of the localization of food industries. We use a spatial econometric model to account for spatial dependence and agglomeration and isolate the impact of economics, regional factors and the French Policy dedicated to clusters *(pôle de Compétitivité)*.

2 Location of Agri-Food Industries and Competitive Clusters

2.1 *Location of Agri-Food Industries and Spatial Dynamics*

According to the localisation map, showing prevalence in 3883 districts (canton) around metropolitan France, food industries are spread throughout the land especially in larger cities (Paris, Lyon, Toulouse, etc; Fig. 18.1a and b). However, some concentrations of food industry stand out. They are mostly in the West of France, notably in the regions of Brittany and *Pays de la Loire,* but also found in the East, South and South-East *(Languedoc-Roussillon,* etc.).

Beyond these observations about localisation, the rest of the analysis describes the spatial autocorrelations, positive and negative, resulting from the presence of such food industries. Anselin defines 'spatial autocorrelation' as *"a measurement of the incidence of similar values and similar location"* (Anselin 2001). So, when the spatial autocorrelation is positive, this indicates that the variable doesn't follow a random distribution and is surrounded by other values of the same size (ex. A weak, random value would be associated with other weak values). Positive, spatial autocorrelation indicates the agglomeration phenomenon. It is said to be negative when the values around the random value are not the same. The Open GeoDa software, allows these spread charts to be studied using LISA tests of spatial association. Data

Fig. 18.1 (**a**) Food industries location in 2000, (**b**) food industries location in 2007

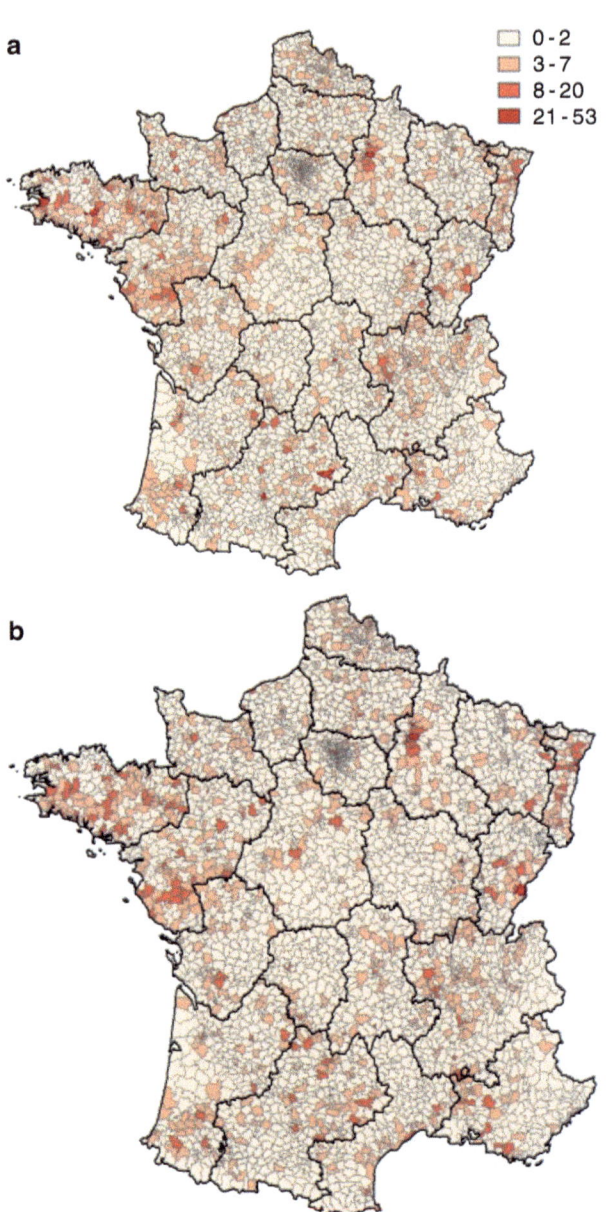

was collected at (canton) district level. However, the mapped results at the employment zones[1] showed pertinence to the problematic. Indeed, Industrial/political decisions are mostly taken at this level. This is the same for a number of pertinent indicators in our study (employment, competitivity zone…) giving spatial apportionment and significant autocorrelation. We have chosen to present the results from both of these geographic levels (Fig. 18.2).

High-High (HH) and Low-Low (LL) correspond to spatial association of similar values (Anselin 1995). So red areas, high concentration (of food industries), or blue areas, low concentration, are surrounded by neighbouring areas with similar concentrations. The term 'Neighbouring' areas refers to the nearest (adjacency 1). So, high-low and low-high, indicated by zones of light colour, show spatial association of unrelated values.

On the map showing spatial agglomeration of 5 % level of significant, agglomeration can be observed around Brittany and, to a lesser extent, the Parisian region and South of France. Areas presenting weak concentration of agribusiness, in blue, are also shown (centre and South East). Light-coloured areas (HL-LH) show atypical zones (mainly central Brittany). These atypical zones are more widely dispersed, where industries have setup within a work zone with few neighboring companies. The study of other factors could therefore be useful in exploring these exceptions.

2.2 Location of Competitive Clusters with Agri-Food Vocation

The only competitive clusters considered in the analysis are those subject to MAAP (the French ministry of agriculture, food, fisheries, rural life and land use planning) that carry out an activity concerned with agri-food and/or agriculture and are strongly tied to the territory in which that activity is conducted.

Clusters related to sea food products were not included, nor were the InnoViandes and Prod'Innov clusters, as in 2010 they lost their national accreditation as 'competitive cluster'. Finally, biotechnology, health and nutrition clusters were not looked at because they are highly R&D-oriented and less directly linked to agriculture and agri-food sectors. The following clusters were studied: *Industrie et Agro-ressources* (IAR), *Nutrition Santé Longévité* (NSL) and Vitagora. All clusters mentioned will however be taken into account in the last part of the study – network analysis – since all of them collaborate with agricultural or agri-food competitive clusters.

According to DATAR (*Délégation interministérielle à l'aménagement du territoire et l'attractivité régionale*) in charge of spatial planning and regional beauty), 14 clusters are subject to MAAP, of which 6 meet our criteria. Also, the following clusters are selected (Table 18.1).

The localisation of these clusters is presented on the following map, with their headquarters and the five work zones which regroup the greatest number of adherents

---

[1] An 'employment zone' is defined by the national institute of statistical economics information (INSEE) as "a geographical area within which a majority of employees live and work, and within which establishments can find the necessary labour to perform commissioned work. The definition of employment zones creates areas adapted to local research of the work market and mark out areas for use by local or public authorities.

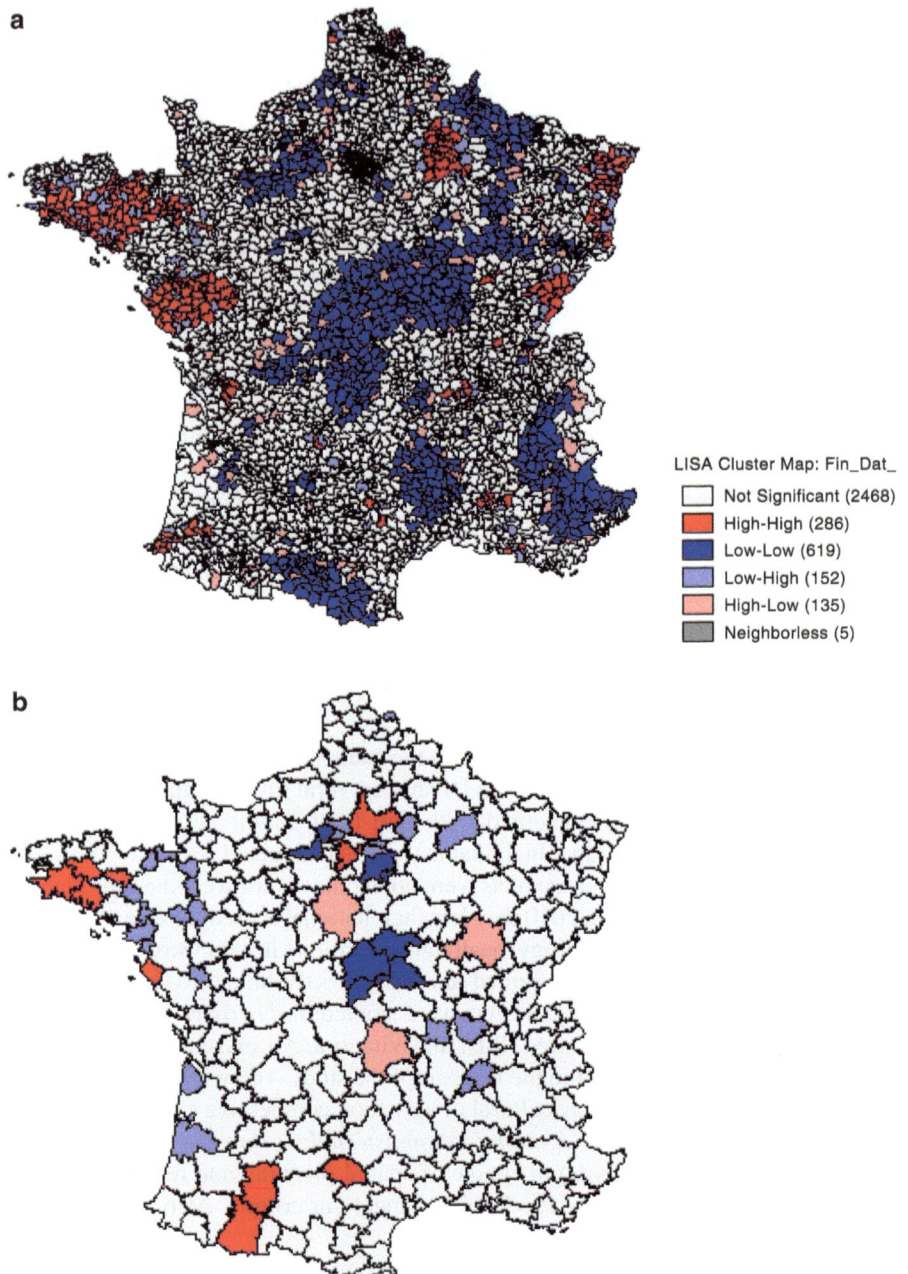

Fig. 18.2 (a) LISA of food industries in 2007 at the 'Canton' level, (b) LISA of food industries in 2007 at the "Zone d'emploi" level

18 Main Factors Affecting Food Industry Clustering in France

Table 18.1 Competitiveness clusters studied

Name of cluster	Location of cluster headquarters	Main issues dealt with
Agrimip Innovation	Castanet Tolosan, Midi-Pyrénées	Food supply chains
Céréales Vallée	Saint-Beauzire, Auvergne	Cereals
Pôle Européen Innovation Fruits et Légumes	Avignon, Provence-Alpes-Côte-d'Azur	Fruits and vegetables
Qualiméditerranée	Montpellier, Languedoc-Roussillon	Fruits and vegetables, wine-growing, cereals and Mediterranean crops
Valorial	Rennes, Bretagne	Foods of tomorrow, milk, meat and egg products, agri-food technologies and nutrition
Végépolys	Angers, Pays-de-la-Loire	Varietal selection, horticulture, specialised plants, landscapes, wine-growing, market gardening

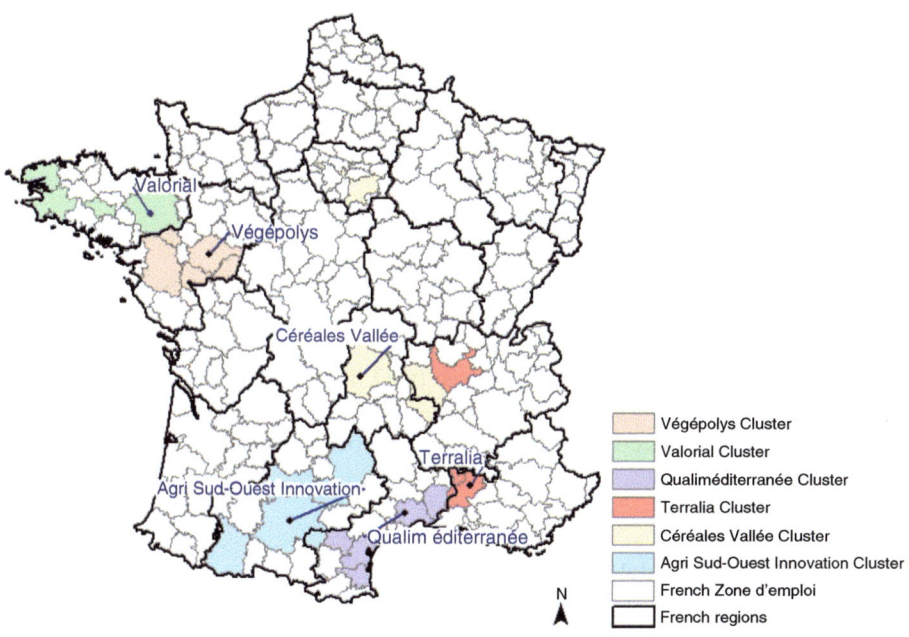

Fig. 18.3 Location of agri-food cluster in France

(number of plants). This map shows that certain headquarters are found at the heart of the dynamic of the adherent companies (Valorial, Vegepolys, Agri Sud Ouest Innovation, Qualiméditerranée), while for others this proximity is less pronounced (Terralia, Céréales Vallée). In the following part, the potential effects of this clustering on agri-food activity will be explored (Fig. 18.3).

3 Spatial Regression Models

3.1 *Empirical Model*

Consideration of spatial autocorrelation in econometric models can be done in several ways: by a spatial autocorrelation at lag with endogenous or exogenous variable offset by a spatial autocorrelation of errors, or autocorrelation containing both an offset variable and autocorrelation of errors.

3.1.1 Spatial Lag Autocorrelation

In the spatial auto-regressive model (SAR), a "lagged endogenous variable" Wy is included in the classical linear regression model.

$$y = \rho Wy + \beta X + \mu$$

Wy is the lagged endogenous variable for the weight matrix W, ρ is the spatial autoregressive parameter indicating the intensity of the interaction between observations y. In this model, the observation y_i is partly explained by the values taken by y in neighboring regions.

3.1.2 Spatial Error Autocorrelation

To specify spatial error model (SEM) autocorrelation, an auto-regressive process is usually used on the errors:

$$y = \beta X + \mu, \text{ avec } \mu = \lambda W \mu + \varepsilon$$

The λ parameter reflects the intensity of interdependence between residues and u is the error term.

3.1.3 General Spatial Autocorrelation

Formally, the general spatial model (SARMA) is a combination of the two previous models.

$$y = \rho Wy + \beta X + \mu, \text{ avec } \mu = \lambda W \mu + \varepsilon$$

The resolution of these models requires the use of specific software that can take into account the spatial dimension and relies on Geographic Information System software (SIG).

In this paper, we use a Spatially Weighted two Stage Least Squares estimation (spatial instrumental variable IV estimation), suggested by Kelejian and Prucha (2007), which enables us to analyze both types of endogeneity (spatial lag and Cluster).[2] This approach consists of using lower orders of the spatial lags of the exogenous variables as instruments for the endogenous spatial lag, together with other instruments for other endogenous variables.

We first estimate the model using an ordinary least squares (OLS) regression. The results are scrutinized for the existence of spatial patterns based on the Lagrange multiplier (LM) principle to test against spatial error or spatial lag alternative (Anselin and Bera 1998) using the Geoda software. The robust LM statistics for spatial lag and error dependence are found to be significant, clearly indicating a general spatial model SARMA.

3.2 Variable Description

The objective of this article is to highlight the factors that explain the location of food industries, and to identify the role of competitive clusters. To do so, we use a spatial econometric model to account for spatial dependence and agglomeration and to isolate the impact of economics and regional factors. Data are mainly related to local infrastructure (roads and subways, harbors, airports...), urbanization (urban areas, agricultural areas), employment (number employees, unemployment rate...), population (density and population), other industries (business and transport), and public policies (Cluster policies).

Working on these issues at a regional level, it is not possible to precisely determine the impact of business clusters. Indeed, the clusters are a new industrial policy, intended to boost employment and the competitiveness of industries which, in a knowledge economy, have to be innovative. Data about employment and industry in France are generally shared across 'employment areas'. We have therefore chosen to perform the regression at 'employment areas' level. The dependent variable considered in this analysis is the number of Food Processing plants with more than 20 employees in each employment area (y). To introduce spatial dependence and take into account the spatial nature of the data, we integrated a spatially lagged variable (the number of agribusinesses in neighboring areas of employment). For this we created a weight matrix (neighborhood). Data on the number of industries is given by the statistical foresight service (SSP) of the Ministry of Agriculture, Annual Survey of business EAE. The latest available data are from 2007.

---

[2] We first estimated the model using an ordinary least squares (OLS) regression. The results were scrutinized for the existence of spatial patterns based on the Lagrange multiplier (LM) principle to test against spatial error or spatial lag alternative (Anselin and Bera 1998) using the Geoda software. The robust LM statistics for spatial lag and error dependence were found to be significant, clearly indicating a general spatial model SARMA.

To study the proximity of food industries with the agricultural sector, we integrated the share of agricultural employment in total employment (source INSEE). This variable can be regarded as a "Proxy" local agricultural dynamics. Indeed, agricultural employment provides information on the distribution of farms and companies offering associated services. As a source of raw material, the agricultural sector seems at first sight to be an interesting indicator explaining the distribution of agribusinesses.

In the Food industry, 52 % of companies have more than 250 employees while almost no companies have fewer than 10 employees. The agribusinesses are then, a major source of employment. This means they need to locate in areas with high potential assets. In order to take the link between industry and the presence of labor into account, we used the number of unemployed workers (the "unemployment" variable). We have also included the variable "median wage household" as an indicator of wealth and attractiveness of labor.

To study the proximity of food industries with downstream firms we considered the number of commercial outlets. To study French territorial land, and in particular the impact of urbanization, we used the land surface artificialized (source: Corine Land Cover) (surface of land dedicated to urban construction).

The area of artificial territories illustrates trends of urbanization for different employment zones. We also considered the population density as an indicator of urbanization and urban spatial dynamics. Policy implementation of competitiveness clusters is clearly based on the concept of Clusters as defined by Porter (1998) *"Clusters are geographic concentrations of interconnected companies and institutions in a Particular field"* to promote innovation, value added vector and employment within a territory. The competitive clusters are therefore based on existing industrial dynamics.

For clusters of food industries, the link with local industrial dynamics is not obvious in France and this is different depending on poles and sectors (Ben Arfa et al. 2009). To see if the competitive clusters have an impact on the location and agglomeration of agribusiness, we introduced a 'Cluster' variable to the model. It takes into account the number of firms located in the 'top five employment areas' which are members of one of the food industry clusters. Because data about competitive clusters is very limited and geographic level is the employment area.

This variable 'cluster' was introduced as an endogenous variable in the model as it is believed that the implementation of clusters is related to the location of industries, but industries can be influenced by the presence of competitive clusters. The local transport infrastructure, as measured by the road network within the area of employment, as well as the number of young higher education degree holders, which may be an indicator of the link between research and training and the presence of competitive clusters, are used as instrumental variables for clusters.

We have chosen to use an adjacency matrix, to study the phenomena of possible distribution between employment areas, and to see the impact of competitive clusters on the location of agri-food industries.

4 Results and Discussion

The application of the employment area model firstly shows that spatial autocorrelation is negative (rho = −0.49). This result suggests that, at this geographical level, the effects of competition exist between neighboring areas, and that generally no effects exist of the agglomeration beyond the employment zone. This can be explained by the very definition of employment areas which are determined by the economic activity that develops within them.

Map (Fig. 18.2b) shown in the first section, can indeed demonstrate this result by noting that there are ultimately more atypical areas with high-low or low-high combinations, that is to say areas with a high number of industries surrounded by neighbours with low numbers, or the inverse. Model results are shown in Table 18.2. Beyond spatial autocorrelation, analysis of the factors explaining the spatial dynamics of agri-food industries shows that agricultural employment has a significantly positive impact on the location of the food industries. Indeed, food industries are located closest to the supply areas, where agricultural growth is the highest, which creates greater agricultural employment. Another interesting result is that industries locate in more urban areas with a high population density. This can be explained by access to urban services necessary for the development of industries but which also attract more workers and consumers. These two results confirm that the location of

Table 18.2 Spatial regression of food industries in 2007[a]

Variable	Coefficient	Std. error	z-Statistic	Probability
Constant	8.9397598	14.3369286	0.6235478	0.5329246
Population density	0.0280322	0.0048059	5.8328744	5.45E-09
Median wages	0.0003453	0.0008136	0.4243324	0.6713234
Share of agricultural employment	1.237152	0.659215	1.8767049	0.06055855
Unemployment share	−0.0003921	0.0003754	−1.0446542	0.2961828
Number of commercial establishments	0.0003267	0.0003322	0.9835618	0.325331
Urban area	0.0010694	0.0004358	2.4540484	0.0141258
Clusters	1.8171095	0.7411954	2.451593	0.014223
Rho	−0.4908918	0.1263008	−3.8866869	0.000102
lambda	0.4501493	0.1501273	2.9984503	0.00271
Pseudo- R^2	0.543164			
Spatial pseudo- R^2	0.580135			

[a]The significance of the parameter 'LAMDA' expresses a nuisance in model specification indicating (i) either structural data dependency, where we see a structural distribution of industries, with more industry in peripheral areas of France (Grand West, Alsace around the Mediterranean) and much less in the, rather rural, centre of France (ii) or an omission of significant variable explaining the location of industries not included in the model (sectoral effects within the food industry branch)

food industries springs from access to both agricultural inputs and a market (home market effect). Finally, it should be noted that the variable relating to the representation of competitive clusters in areas of employment is significant and positive. This result suggests that the policy reinforces the food industry agglomeration dynamic. Specifically, a high level of adherence to the food industry in competitive clusters enhances the effect of agglomeration, itself a competitive factor.

5 Conclusion

In this article, we analyse the phenomena of agglomeration among food industries. We seek to determine to which extent the French led policy of regional economic development, the establishment of 'competitive clusters', contributes to this agglomeration, among other factors.

Analysis of the spatial autocorrelation of food industries, has allowed us to highlight the presence of (canton) district level agglomeration (small administrative level), however this phenomenon is not the same if it is observed in employment areas. Indeed, at this scale, an effect of competition about attractiveness between areas of employment can be observed. Meanwhile, we observe that the 'competitive clusters' dedicated to the agro-food sector are located in the West and South of France and their HQs are in relatively heterogeneous areas from the point of the observed spatial dynamics of food industries. Only Valorial (Brittany) and Végépolys (Pays de la Loire) are in an agglomeration zone of food industries. Beyond this, we analyse to which extent the influence of the poles on the territory strengthens, along with other factors, the agglomeration of activities, which is a factor of competitiveness.

Thus, we apply a spatial econometric model that confirms that the process of spatial autocorrelation is negative in employment areas. However, traditional factors of geographic concentration of activities play in a conventional manner (agricultural employment, infrastructure, urbanisation) and it is interesting to note that the influence of 'competitiveness clusters' increases agglomeration. This analysis constitutes an original work which could be taken further. While, it would be interesting to compare the application of this model to different geographical models, even for an analysis at district level, the issue of data availability is a real constraint. Moreover, it would be interesting to test new variables, including those that would reveal sectoral effects. Finally, when the policy of competitiveness cluster setup has gained experience, it will be interesting to analyse whether this model is stable, or if its efficiency increases, over time.

References

Anselin L (1995) Local indicator of spatial association-LISA. Geogr Anal 27:93–115
Anselin L (2001) Spatial econometrics. In: Baltagi B (ed) A companion to theoretical econometrics. Basil Blackwell, Oxford, pp 310–330

Anselin L, Bera AK (1998) Spatial dependence in linear regression models with an introduction to spatial econometrics. In: Ullah A, Gilles DA (eds) Handbook of applied economics statistics. Marcel Dekker, New York, pp 237–289

Bagoulla C, Chevassus E, Daniel K, Gaigne C (2010) Trade liberalization and regional location: evidence from the French agroindustry. Am J Agric Econ 92(4):1040–1050

Ben Arfa N, Rodriguez C, Daniel K (2009) Dynamiques spatiales de la production agricole en France. Revue d'Economie Régionale et Urbaine 4:807–834

Ben Arfa N, Daniel K, Fontaine F (2013) Agri-food clusters: is French policy in line with real spatial dynamics? Région et Développement 38:249–279

Duranton G, Martin P, Mayer T, Mayneris F (2010) The economics of clusters, lessons from the French experience. Oxford University Press, Oxford

Kelejian HH, Prucha IR (2007) HAC estimation in a spatial framework. J Econ 140(1):131–154

Leslie SW, Kargon RH (1996) Selling Silicon Valley: Frederick Terman's model for regional advantage. Bus Hist Rev 70(4):435–472

Porter ME (1998) Clusters and the new economics of competition. Harv Bus Rev 76(6):77–90

Chapter 19
Industrial Cluster Analysis, Entrepreneurship and Regional Economic Development

Roger R. Stough and Junbo Yu

Abstract This paper clarifies the cluster concept and how clusters are identified, nurtured and grown as elements of regional economies, with a special focus on food specific clusters.

It points out that based on analogous examples and theoretical arguments that Northeast Asia appears to be well positioned from a geographic perspective to develop a world class food industry. However, for considering the competitiveness of Northeast Asia, several questions are raised: to what extent will conventional geographic factors function amid the globally restructured food industry and how are these geographic factors going to affect the future formation, development and dynamics of food industry clusters across the world? To answer these questions analytic and visualization techniques including the MSA matrix and Input-output modeling are considered for the identification of clusters and related development planning and implementation.

A qualitative analysis of the institutional framework and support elements that exist (or do not) in a region is also provided. This analysis focuses on the concept of smart infrastructure. Three cases in Europe: the Oresund Food cluster in Denmark and Sweden; Greenport in South Holland and the Dutch Food Valley in the eastern part of the Netherlands in the lower Rhine River valley are examined and discussed to illustrate the institutional relationships and smart infrastructure of advanced or internationally competitive food clusters.

Keywords Northeast Asia • MSA matrix • Input-output modeling • Smart infrastructure

---

R.R. Stough (✉)
George Mason University, Fairfax, VA, USA
e-mail: rstough@gmu.edu

J. Yu
School of Administration, Jilian University, Jilian, China

1 Introduction

This paper examines the industrial cluster concept and how clusters are identified, nurtured and grown as elements of regional economies, with a special focus on food specific clusters. There is a vast literature on clusters that for the most part argues that they complement and support successful economic development and growth. Despite the recent strong and growing interest around the cluster concept it has been around for a long time (see Marshall 1920; Perroux 1950; Porter 1991) and thus the body of cluster research has quite a "long tail". Nonetheless, the economic significance of the cluster concept lies in the conclusion by theorists and many (but not all) empirical researchers that clusters amplify economic performance because they facilitate development and maintenance of economies of scope and scale in regions.

Regional industrial clusters have been defined by many including Information Design Associates and ICF/Kaiser International, Inc. (1997) which we rely on in this paper:

> ......Industry clusters are agglomerations of competing and collaborating industries in a region networked into horizontal and vertical relationships involving strong common buyer-supplier linkages, and relying on shared formulations of specialized economic institutions. Because they are built around core export oriented firms, industry clusters bring new wealth into a region and help drive the region's economic growth (p. 3).

Despite the great interest in the cluster concept both conceptually and operationally it is noteworthy that processes of knowledge production and its conversion into economically useful knowledge are paramount. Thus it is essential to consider the knowledge conversion processes and infrastructure needed to support dynamically evolving and adjusting industrial clusters.

The title of this paper signifies the critical relationship between entrepreneurship and economic development in the context of cluster centric regional economic development. The literature on economic development and entrepreneurship clearly supports a conclusion that the two are positively and strongly related at almost all levels, i.e., from the micro or firm level to the national and in fact the transnational (Rocha 2004). While it is believed or perhaps assumed and broadly argued that industrial clusters contribute positively to economic development and entrepreneurship (and vice versa) the evidence is more complicated and in fact, in part, ambiguous (Rocha 2004). Despite this lingering of mixed results, in this paper we will assume that the conceptual arguments about the contribution of clusters to economic development based on agglomeration and interactive positive spillover economies are sufficient to undergird the assumption that all three are positively related.

It is argued that we are living in a knowledge age and one where knowledge is becoming the critical input to successful economic growth and development and cites (Audretsch et al. 2006). From this follows an assumption or conclusion that the presence of a university, as a primary knowledge creating institution in a region will contribute fundamentally to the competitiveness of its economy. While there seems to be some evidence to support this conclusion (Hart 2003) it is not at all clear that having institutions that are major producers of knowledge in a region will ensure the creation of greater productivity and economic growth. Most importantly, this is because the raw or modestly refined knowledge produced in the typical academy institution is just that. Without significant development and enhancement of raw knowledge there will, at best, be marginal development of economically useful knowledge, i.e., commercially valuable products and services.[1]

There are barriers that face the conversion of raw knowledge into useful knowledge. These range from the simple issue of recognizing knowledge that has potential economic value to a host of formal and informal institutional barriers that constrain the conversion process (or abet it) such as capital formation, incubation, supporting organizations/associations, availability of advanced professional services locally and generally supportive institutions including the regulatory environment. Acs et al. (2006, 2004) argue raw knowledge must pass through a filter (the knowledge filter) on its journey to becoming economically useful. Some regions have a course-grained, i.e. highly porous, filter and thus the conversion of knowledge into economically useful knowledge will be relatively high and thus facilitated; regions with a non-porous filter will have less success because the supporting institutions and regulations present more barriers to commercialization. Acs et al. (2004, 2006) also argue that entrepreneurs are the agents that primarily transform knowledge into economically useful knowledge and thus "drive" raw knowledge through the filter. In short, entrepreneurs and their supporting institutions are equally important inputs to high levels of economic performance.

2 Elements of the Competitiveness of Northeast Asia

Unlike many other industries that are swiftly becoming footloose and concentrated in the knowledge economy, the food industry in general remains relatively bounded by geography and thus notably multi-centered with hundreds of recognized food clusters.[2] However, incremental, yet profound changes, have already taken place in most of the world-class food industry clusters in developed countries that are featured by such important characteristics as switching from bulk production to functional food, convenience food and food service, significant increases in R&D investment, the formation of a sectoral innovation system surrounding the cluster, and the expansion of the "breadth" and "depth" of the cluster (Lagnevik et al. 2003) in both an organizational and geographic sense. While the occurrence and reinforcement of these changes will constantly reshape the landscape of the global food industry in the next few years, an intriguing question is: how will conventional geographic factors be positioned amid the globally restructured food industry and

---

[1] This is not to argue that universities are unsuccessful in efforts to produce economic useful knowledge in the form of patents and licenses and new company formation. For sure there are highly successful cases such as MIT and Stanford and others that are improving their performance. For the most part however universities have had marginal success to date.

[2] See a list of world-class food industry clusters identified by researchers at http://www.isc.hbs.edu/MetaStudy2002Bib.pdf.

how these geographic factors will interact with the aforementioned "changes" to affect the future formation, development and dynamics of the food industry clusters across the world?

2.1 Geography Matters

From Marshall (1920) to Krugman (1995), economists have persistently endeavored to convince the general public and policymakers that "geography matters", particularly in the process of industry agglomeration and trade – the two most decisive mechanisms that enable the formation and development industry clusters. Unlike professional geographers, an economist's interpretation and concern regarding the term "geography" are usually confined to (1) physical accessibility, (2) natural endowments that associate with the location and, (3) historical incidents that were tied to the location. Accordingly, in any attempt to explain the formation and development of a particular industry cluster, economists would typically start by highlighting its geographic features from these three perspectives and proceed by stressing their causal relationship with the "Holy Grail" of economic growth – economies of scale and increasing returns.

In view of this, the geographic explanation of food industry cluster formation and development in the existing economic literature offers no exception. First, physical proximity to the primary resources and food products market, which is often taken as a proxy for accessibility, has been widely believed to be the most conducive and important geographic factor. From an economic perspective, primary resources and a capacious market respectively stand for the supply and demand factor and the opportunity to connect them by fostering a food industry that can process primary resources in accordance to the requirements of the consumers in the market. Proximity to either the primary resources or the food product markets or both of them would therefore reduce the cost of linking the supply and demand and therefore put more leverage on impacting them effectively. However, it is also found that, in contrast to the production of primary resources which is fairly stable and incremental from the perspective of product innovation and technological change, the food product market normally exhibits a high degree of volatility due to the constant changes and upgrading that occur in consumer preferences. As a consequence, food industry clusters that are adjacent or imbedded in their major markets tend to outperform their peers that merely stay close to primary resources in the long-run because they have been better positioned to detect and react to new trends in the industry.

2.2 What Else Matters?

A well-known countering argument for the preceding argument is the California wine industry that is situated together with its grape producers. Technically, the success of this location seems to be caused by the difficulty of preserving and transporting grapes and thus makes transforming them into wine first and then bottling and delivering it to the customers a more economical solution. However, a retrospect to the history of American wine industry would inspire people to raise the following question: if wine producers were technologically confined to stay close to local grape farms, why or how did California eventually prosper and supersede others to become a national brand? An elementary answer to start with is the state's unique climate associated with its unique location as an indispensable natural endowment to yield perfect grapes qualified for wine production. While many scholars insist that the success of the California wine industry should be attributed to the formation of an industry chain centered on the wine industry itself, e.g. packaging, production machinery, distribution, warehousing and quality control, the source of these supportive industries' confidence and devotion would nevertheless remain a mystery without taking into account California's natural endowments.

Historical incident is another crucial factor that features the success of some other world-class food industry clusters. For instance, the Oresund food cluster located in Demark and Scania in southern Sweden used to be a regional industry cluster serving only two nations before the end of the Cold War. However, the formation of the Single European Market and the collapse of the Soviet Union since the early 1990s presented the Oresund food cluster an unprecedented market with easy accessibility, culture resemblance and most importantly, a voluminous, sustainable and diversified demand. Today, the Oresund cluster has become the most noticeable and fastest growing one among its peers. Its natural and traditional market has expanded in northern Europe from the UK to Russia and it dominates the Scandinavian food market. Due to the high concentration of retailers and wholesalers in the area and the well-developed distribution system, a company located in the Oresund area can, by working with a handful of customers and supporting companies, serve almost all of the 23 million consumers in Scandinavia and the 100–200 million in the Baltic rim. In addition, the Baltic States are expected to continuously offer good market potential and an expanding food market along with rising living standards and changing consumer patterns.

2.3 Lessons for Northeast Asia

The answer to the question – where, when and how can people in Northeast Asia establish world class food industry clusters is informed by our previous discussion on the geographic determinants of food industry cluster. Northeast Asia, when viewed as a whole, offers great accessibility to Eurasia, the rest of the mainland China, Southeast Asia, Oceania and the entire America continent; "domestically", it is the one of the most populated areas in the world where Japan and South Korea are among the most developed counties with high income populations, Northeast China and Siberian Russia are fast growing and upgrading, and North Korea is near the dawn of socioeconomic reform. This region is also blessed with abundant natural endowments for food industry development where, for instance, the best quality of rice and marine products can be found in Chinese provinces like Ji-lin, Hei-longjiang and Liao-ning, in the Korean peninsula and in Japan; and, most importantly, the prospect of an East Asia Regionalism (Dent 2008) could create a historical opportunity to achieve unprecedented economic integration within the Northeast Asia region for reducing transaction costs, increasing resource allocation efficiency and enhancing overall competitiveness. Therefore, based on analogous examples and the theoretical arguments Northeast Asia appears to be well positioned from a geographic perspective to develop a world class food industry.

3 Identifying Clusters and Their Dynamics for Economic Development and Growth: Quantitative and Qualitative Analyses Used to Promote and Support Cluster Development

There are many scholars and planners who have investigated industrial clusters and theorized (and in some cases conducted empirical research) about their nature as proposed by the cited definition above. In this context, Porter (1991) identifies four elements that are critically important for identifying and integrating relevant economic information and forming a strategy for planning and growing clusters (and regional economic growth and development). Once clusters are identified integrating these elements into a strategy is necessary. These elements are (see Porter 1991):

1. Demand conditions;
2. Firm strategy/rivalry;
3. Related and supporting industries; and,
4. Factor conditions.

Data is necessary to assess each of these and to support combining them into a scheme or strategy for growth and development. For example, existing data can be used to determine if clusters (and industry sectors that make them up) are growing or contracting and/or for estimating their potential. Data based estimates can be used to examine competition in an international context and thus the level of international competition facing a cluster (Fig. 19.1) and more specifically to estimate whether clusters are in markets where there is increased global competition. Further, using interviews and focus groups, specific issues that planners may consider foundation elements in planning for cluster development and more generally economic development and planning, can be examined such as jobs, infrastructure, quality of life, housing institutional and regulatory quality and social and professional networks that may exist, at least in part, to promote cluster success can be identified in the same way. Such techniques can also be used to identify institutional and organizational infrastructure that may provide specialized support.

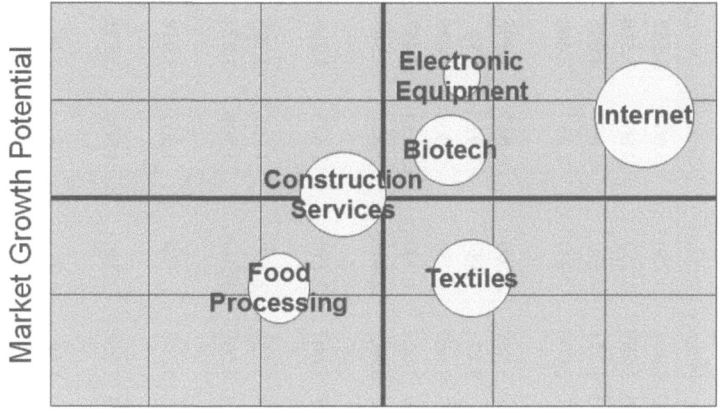

Fig. 19.1 Competitive advantage

Fig. 19.2 The MSA matrix (an illustration)

		Industry Sectors			
		A	B	C	Σ
Evaluation Criteria	X	0	1	0	1
	Y	1	3	0	4
	Z	0	0	2	2
	Σ	1	4	2	

3.1 Analysis Techniques to Support Identification of Clusters and Related Planning

Figure 19.2 presents a Multi-Sector Analysis (MSA) framework in the form of a 2 by 2 matrix with clusters/sectors for the column headings and evaluation criteria for the rows. A fleshed out full MSA matrix for the data collected for a study in the Northern Virginia part of the U.S. National Capital Region (NCR) is presented in Table 19.1 for illustration purposes. The data evaluation ratings on a 5-point scale were collected from groups of CEOs representing companies from each of the 11 clusters. There are seven specific evaluation categories with multiple elements for each category for this study: Transportation infrastructure, finance, human resources, technology and development, international trade orientation, government support, regional economic strengths and some firm specific management factors (strength and weaknesses).

To illustrate how the investigator and planner or policy maker can develop hypotheses, note that scores for the professional services and associations, two

Table 19.1 The MSA tool illustrates responses to how factors affect the performance of industries/clusters in the Northern Virginia region

Competitiveness factors	Sectors										Mean across factors
	Aerospace	Biotech	Information technology	System integrator	Telecomm	Transport	Association	Real estate	Finance	Professional services	Tourism	
Adequate highway system	3.429	4.000	2.143	3.542	3.625	3.091	3.556	4.500	3.444	3.750	3.667	3.552
Scheduled air service	4.000	4.375	2.571	3.292	3.500	3.182	4.000	3.250	2.889	4.250	4.667	3.510
Telecommunications	4.571	4.625	4.429	4.375	4.375	3.546	4.000	3.250	4.111	4.250	3.667	4.095
Environmental and waste mgmt.	2.570	3.000	1.857	2.250	2.625	1.818	2.667	3.167	2.000	2.000	2.333	2.400
Regional quality of life	4.143	4.125	3.571	3.958	4.500	3.273	3.444	4.417	3.889	3.500	3.833	3.905
Finance	3.00	4.31	2.64	2.81	3.19	2.09	2.28	2.96	3.17	1.38	1.92	2.78
Availability of financing	3.143	4.375	3.000	3.125	3.625	2.455	2.556	3.417	3.667	1.750	2.000	3.095
Venture capital	2.857	4.250	2.286	2.500	2.750	1.727	2.000	2.500	2.667	1.000	1.833	2.457
Human resource development	4.00	3.45	3.49	3.94	4.18	2.45	2.98	3.25	3.62	2.80	3.87	3.50
Higher education/training services	4.000	3.375	3.286	3.958	4.125	2.455	2.778	2.333	3.667	2.250	2.833	3.286
Availability of skilled labor	3.571	3.500	4.000	4.375	4.500	2.182	3.222	3.833	3.556	2.750	4.167	3.705
Availability of prof. employees	4.429	4.000	4.429	4.500	4.375	2.909	3.667	3.250	4.000	3.500	3.500	3.924
Flexible labor-mgmt. relations	4.000	2.750	2.429	2.833	4.000	2.091	2.222	3.250	3.333	2.750	4.333	3.010
Competitive wage/salary structure	4.000	3.625	3.289	4.042	3.875	2.636	3.000	3.583	3.556	2.750	4.500	3.591
Technology and development	2.14	3.15	2.11	2.06	2.63	2.60	1.69	1.70	1.60	0.95	1.83	2.06
University research programs	2.000	2.875	2.429	2.125	2.625	2.273	2.000	1.917	2.000	1.750	1.667	2.162

19 Industrial Cluster Analysis, Entrepreneurship and Regional Economic Development

University-industry partnerships	2.286	3.625	2.429	2.208	3.125	2.636	1.889	1.750	2.333	1.000	2.500	2.352
Federal research lab programs	2.286	3.000	2.429	2.000	2.375	2.182	1.667	1.667	1.111	1.000	2.000	1.971
State research initiatives	1.857	3.125	1.571	1.833	2.250	3.364	1.333	1.583	1.111	0.500	1.667	1.810
Private research efforts	2.286	3.125	1.714	2.125	2.750	2.546	1.556	1.583	1.444	0.500	1.333	2.000
International trade orientation	**3.23**	**2.83**	**2.11**	**2.05**	**2.03**	**1.90**	**1.16**	**2.25**	**2.09**	**0.25**	**2.60**	**2.09**
Current overseas trade activities	3.286	3.375	2.143	2.417	1.750	1.800	1.222	2.333	2.222	0.000	2.667	2.212
Foreign investment into this region	2.429	2.125	1.286	1.625	1.500	1.400	1.222	2.333	2.444	0.500	3.500	1.846
Overseas investment of your firm	2.571	2.250	1.143	0.917	1.750	1.300	0.889	1.833	1.556	0.000	0.500	1.346
Business alliances (w/U.S. firms)	3.857	3.250	3.714	3.250	3.250	3.000	1.333	2.500	2.444	0.500	3.667	2.894
Business alliances (foreign firms)	4.000	3.125	2.286	2.042	1.875	2.000	1.111	2.250	1.778	0.250	2.667	2.144
Government	**3.64**	**3.63**	**3.61**	**3.92**	**3.81**	**2.84**	**3.34**	**4.23**	**3.89**	**3.13**	**4.58**	**3.73**
Local regulation of business	3.429	3.375	3.571	3.917	4.000	3.000	3.444	4.167	3.556	4.000	3.833	3.686
General business climate	3.714	3.875	3.714	4.250	3.875	2.727	3.375	4.417	4.444	2.750	5.000	3.914
Local econ. development efforts	3.429	3.625	3.143	3.250	3.500	2.636	3.000	4.250	3.889	1.750	4.667	3.410
Local tax structure	4.000	3.625	4.000	4.250	3.857	3.000	3.556	4.083	3.667	4.000	4.833	3.904
Regional economic strengths	**3.46**	**2.83**	**3.76**	**3.54**	**3.54**	**3.06**	**2.07**	**3.94**	**3.67**	**2.17**	**4.22**	**3.36**
Performance of your industry sector	3.667	2.875	3.857	4.000	3.625	3.182	2.000	4.000	3.778	2.500	4.833	3.567

(continued)

Table 19.1 (continued)

Competitiveness factors	Sectors										Mean across factors
	Aerospace	Biotech	Information technology	System integrator	Telecomm	Transport	Association	Real estate	Finance	Professional services	Tourism	
Strength of no. VA regional econ.	3.143	2.875	3.857	3.542	3.714	3.091	2.444	4.500	4.222	2.500	4.833	3.558
Cross-industry information flow	3.571	2.750	3.571	3.083	3.286	2.909	1.778	3.333	3.000	1.500	3.000	2.962
Your firm's management characteristics	**3.79**	**3.25**	**3.98**	**3.94**	**3.92**	**3.32**	**3.04**	**3.60**	**4.04**	**2.92**	**3.61**	**3.65**
Customer service/product quality	3.857	3.375	4.571	4.833	4.625	3.818	4.000	4.667	4.889	3.500	4.833	4.381
Inter-business networking	4.143	3.000	4.286	3.708	3.625	3.091	3.000	3.333	4.444	2.500	4.333	3.600
Available management consultants	2.714	2.625	2.857	2.609	2.250	2.818	2.222	2.417	3.000	2.000	1.500	2.519
Marketing capabilities	4.000	3.250	4.000	4.000	4.250	3.455	2.778	4.167	3.889	2.250	4.333	3.762
Entrepreneurship	3.429	3.500	3.857	4.125	4.250	3.182	2.667	4.000	3.889	3.000	3.500	3.686
Info/telecommunication systems	4.571	3.750	4.286	4.375	4.500	3.546	3.556	3.000	4.111	4.250	3.167	3.933
Mean through firm sectors	**3.393**	**3.377**	**3.088**	**3.226**	**3.375**	**2.681**	**2.517**	**3.161**	**3.145**	**2.157**	**3.305**	

important clusters for the NCR, are consistently lower than for the other clusters which suggests that while they are important cluster elements, the respondents see their clusters as being marginal compared to others; this is especially pronounced for technology and development, and international trade. If on the other hand we examine the matrix in terms of the sectors we find considerable agreement across clusters that the region needs improved financial resources, human resource development, higher education resources, and international trade support. The matrix also shows sector specific problem areas such as transportation which appears to have support problems in the areas of environment and waste management that are more severe than for other sectors. Also, aerospace, information technology and systems engineering clusters indicate a concern with the scale and quality of the state's effort to support research and development. As illustrated, the MSA provides evaluation or judgment data that can be used to create hypotheses about the status and/or need for expanded local regional cluster support.

The Northern Virginia region, which is a major component of the U.S. National Capital Region, is a large urban sub-region with 5 counties and a population of about 1.5 million, and some of the highest family incomes in the U.S. over the past 30 years. A smaller region in a less complex economic region might only have one or two clusters.

The MSA can be used to produce several different types of indices. Here in Fig. 19.3 an index of sector competence is illustrated with data from a cluster analysis conducted for the Far North Queensland (FNQ) Area in Australia (Stimson and Stough 2006, pp. 279–318). Large or robust sectors identified for FNQ are, as in Fig. 19.3, food processing, transportation, and tourism and leisure as well as several others including tradable services. Figure 19.4 turns the MSA matrix on its side, so to speak, thus providing a measure of core competencies for the FNQ region: quality of life, market proximity, economic dynamism and good infrastructure. But it is also possible to develop more specialized indices or measures from the MSA such as an export development potential index, Fig. 19.5, where it is shown for the FNQ area that the strong export potential endeavors are agriculture, food processing, tourism and tradable services. It is also possible to develop industry specific risk measures and to identify strong risk factors (price stability, exchange rate and government support). These measures require focus group meetings following construction of the MSA to provide additional insight.

One aspect of cluster analysis is to determine what sectors are directly and/or indirectly related to each other and thereby provide growth opportunities through various promotion efforts at the interface of sectors that are already dependent on each other. One way to estimate the level of economic dependence is through input-output (I-O) analysis (Hewings 1985), which is a methodology for measuring the degree to which sectors in an economy buy and sell goods and services to each other. Figure 19.6 illustrates the form that an inter industry transaction table from and I-O analysis takes. In this table one may identify a potential food and food processing cluster by noting that the links are strong among several industry sectors including agriculture, fishing, energy, wood and paper products, construction, retail services, tradable services and tourism and leisure services. This of course does not

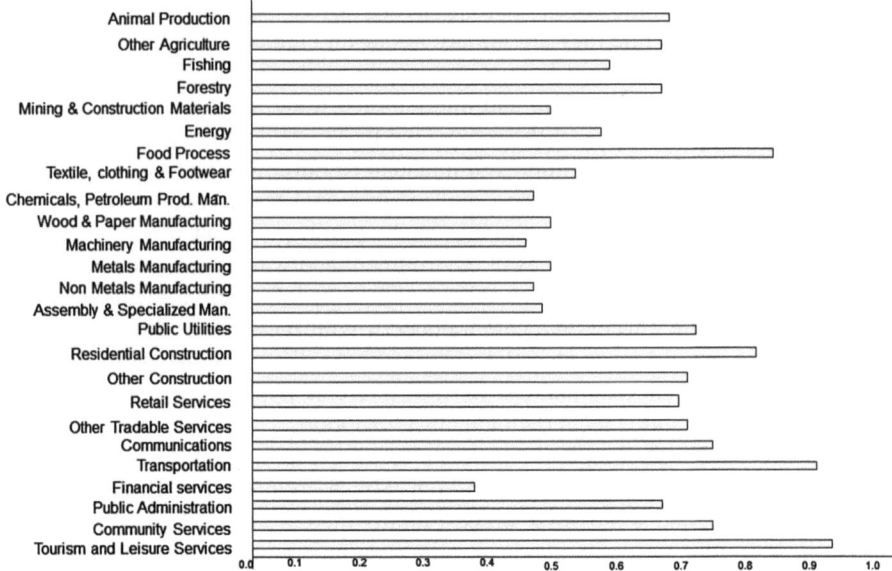

Fig. 19.3 Index of sector industry competence in the FNQ region

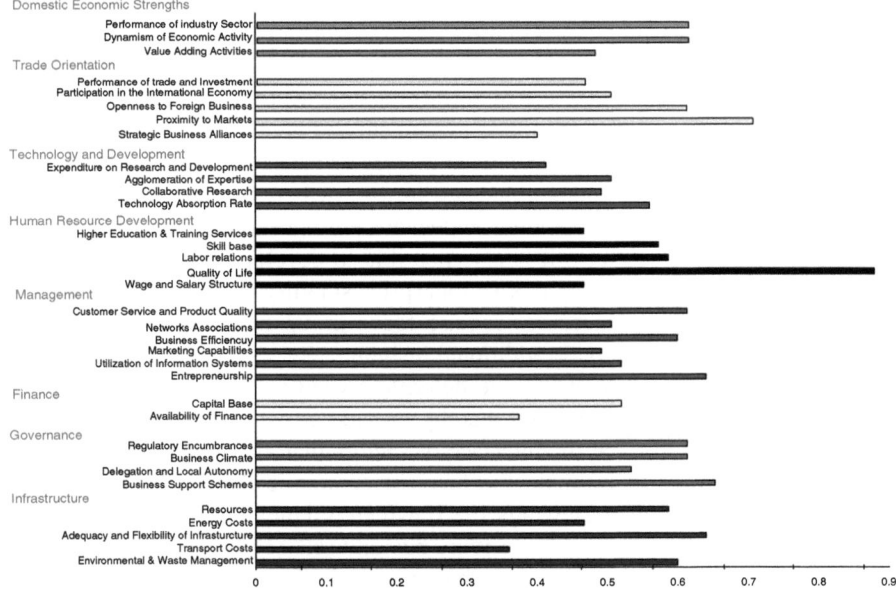

Fig. 19.4 Index of core competencies in the FNQ region

19 Industrial Cluster Analysis, Entrepreneurship and Regional Economic Development   267

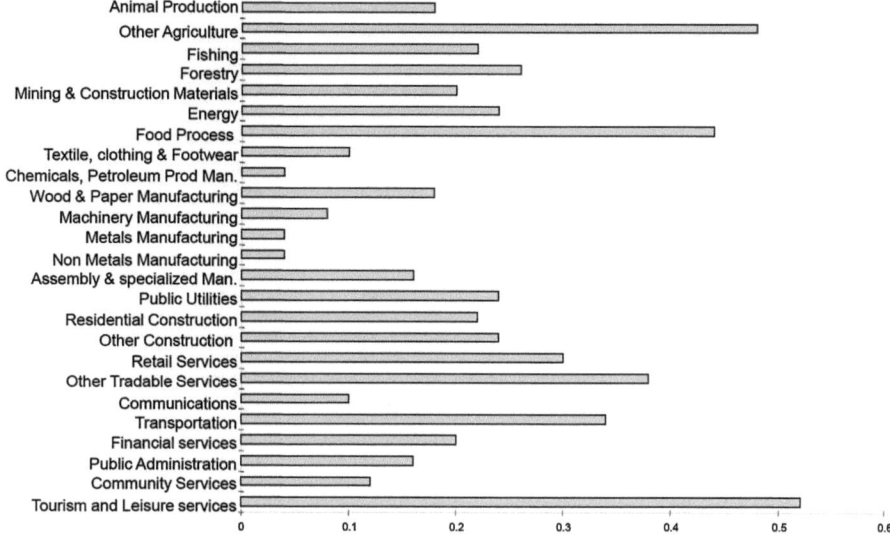

Fig. 19.5 Export industry development potential by industry sector

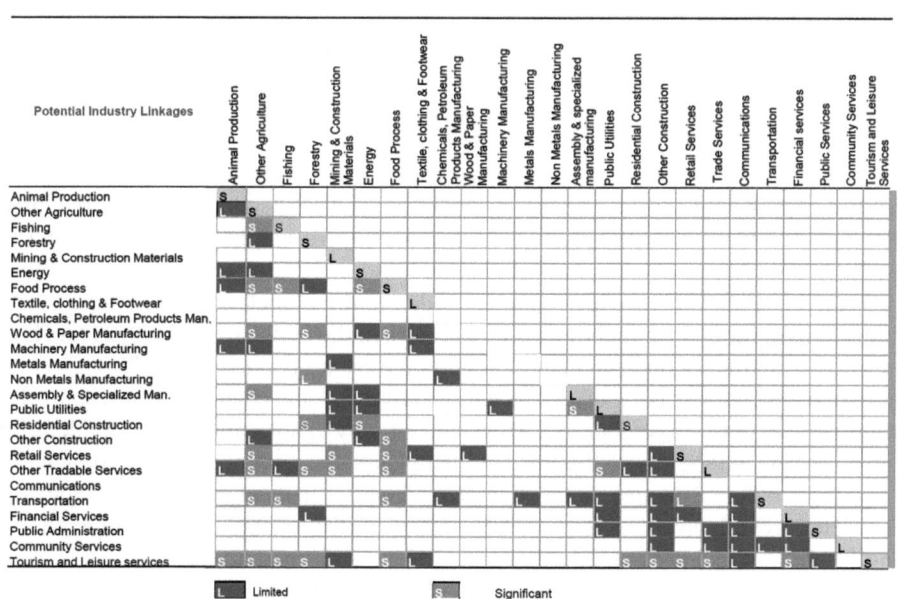

Fig. 19.6 Interactive export industry development potential by industry sector

Rank (in terms of employment in 1993)	SIC	Description	NoVA Employment in 1993	% Change 1988-93	Location Quotient 1988	Location Quotient 1993
	70	Services Total	334,082	18.82%	1.24	1.32
2	8711	Engineering Services	20,434	-17.32%	6.01	4.41
3	8742	Management consulting services	16,671	71.42%	9.35	11.75
5	7371	Computer Programming services	14,696	36.64%	12.90	****
9	8731	Commercial Physical Research	9,864	10.94%	5.85	5.97
10	7373	Computer Integrated Systems design	7,618	-0.35%	****	****
11	7374	Data Processing and Preparation	6,959	8.29%	4.91	4.85
14	8733	Noncommercial research organizatio	5,581	5.18%	4.43	4.17
16	7379	Computer related services, n.e.c.	5,173	156.22%	4.05	8.00
17	8741	Management Services	5,093	16.46%	2.78	2.65
20	8748	Business consulting, n.e.c.	4,772	238.20%	****	8.77
21	7372	Prepackaged Software	4,729	23.05%	****	****
37	7378	Computer maintenance and repair	1,689	61.47%	4.92	6.10
38	8071	Medical Laboratories	1,665	13.65%	1.87	1.85
58	7377	Computer rental and leasing	912	157.63%	4.70	12.46
65	7376	Computer facilities management	714	99.44%	****	****

Fig. 19.7 Employment and location quotient change for selected industry sectors

mean that such a cluster exists or is worthy of further investment as one might focus initially instead on tradable services as a core sector and find that these services which include the same group of sectors also include minerals and mining and public utilities. Planners, analysts and their colleagues and advisors must determine how to structure, define and name the associated cluster on the basis of regional historical knowledge, preferences and experience.

Input-output modeling is the preferred method for conducting industrial cluster analyses however it is time consuming and requires considerable expertise and thus is expensive. Consequently, planners and analysts have sought alternatives that are easier to implement, quicker and yet provide reasonably accurate information. Data presented in Fig. 19.7 provides what is needed to do a quick and for the most part suitable base for cluster analytic work. The rows in the table are industry sectors but at a relatively highly disaggregated level and the columns provide employment and employment change, and industrial location quotients[3] and changes over a relatively short period (in this case 5 years). This data not only enables the analyst to identify large sectors and their growth rates but also their relative importance to some base (e.g., the provincial or state, or the national economy). In most regions and countries this data is readily available and can be used to identify sectors that are large, growing (declining) and are not only important (or not so important) relative to national or provincial/state bench marks. A drawback of this approach over the input-output model is that interdependence must be gauged using qualitative methods such as

---

[3] A location quotient measures the importance of an industry sector in its local economy relative to a reference area economy such as the national economy.

19 Industrial Cluster Analysis, Entrepreneurship and Regional Economic Development 269

focus groups and other information and regional experience rather than determined by data alone. Another is that the time period must be thoughtfully selected to avoid cyclical effects that could bias the results.

Usually when a region feels it needs to undertake cluster analysis and associated planning its economy is already facing stress (see Stimson and Stough (2006) for a discussion on the role of contingencies or stress in regional economic development leadership and planning) and thus an initial "quick study" may be warranted to develop candidate hypotheses to identify possible future directions. This was exactly the situation facing the Northern Virginia Region (NVR), one of the leading knowledge and technology intensive regional economies in the U.S., faced when the first author was asked to conduct such a study in the NVR. The client groups wanted the study completed within 2 weeks! Urgency is part of the business life style in this fast paced and dynamic technology intensive region as the main method of competition is to out-innovate the competition (Stough et al. 1997). Fortunately, a recent input-output analysis existed for the NVR which informed the selection of candidate clusters to be used in producing a MSA matrix (Fig. 19.9). Eleven sectors were identified: aerospace, biotechnology, information technology, system engineering, telecommunications, transport, associations, real estate and construction, finance, professional services and tourism. The rows of the matrix provide the evaluation criteria believed to be important for the study (many of these came from the earlier FNQ study but were restructured for the high technology services economy study of the NVR).

The evaluation data for the MSA were obtained from formal surveys conducted as part of 11 focus group meetings that were distributed in 11 parallel sessions (one session for each cluster grouping) over a catered dinner beginning on Monday evening and ending Thursday evening of the study week. The participants, mostly CEOs of companies representing the 11 clusters, completed the survey for their industry cluster and participated in exercises designed to identify areas they contemplated investing in, i.e., the industry development frontier. The results of the combined analyses of the input-output and MSA research were used to develop illustrations of the core cluster for the NVR and various other but linked clusters as illustrated in Figs. 19.17, 19.18, 19.19, 19.20, 19.21, 19.22, 19.23, 19.24, and 19.25 (see below). The NVR study was completed and presented to the client in less than 2 weeks! Initial planning for a development strategy began in week 3.

The first output of the analysis was the identification of a core cluster and its various components (see Fig. 19.8). The core cluster was composed primarily of large concentrations of advanced technologies largely in the information technology and computer sciences fields along with less intense roles for the electronic chip industry and, the space and bio-sciences. Around this core there were a number of more loosely connected but smaller industry concentrations in areas such as transport and communications, publishing, real estate and construction (especially in the form of a non-local export industry), specialized finance, legal services and manufacturing on or just beyond the far periphery of the region (see Fig. 19.9).

The various smaller clusters are illustrated in Fig. 19.10 – Extra Region Construction and Real Estate, Fig. 19.11 – Information Technology and

Fig. 19.8 Core of the Northern Virginia economy

Fig. 19.9 Core and other important sectors of the Northern Virginia economy

Telecommunications, Fig. 19.12 – Computer Integrated System Design, Fig. 19.13 – Legal Services, Fig. 19.14 – Tourism and Leisure Services, Fig. 19.15 – Finance, and Fig. 19.16 – Transport. The illustrations show the various sub-components or sectors in these clusters.

The results presented above formed the core of the first part of an industrial cluster analysis for the region. The second part focused on how to leverage this knowledge. This was achieved by cluster deepening and cluster stretching. Figures 19.17 and 19.18 illustrate this concept for the Information Technology and Telecommunications (IT&T) cluster in the NVR. The results shown in Fig. 19.18 were developed by posing the deepening question to the focus group for the IT&T

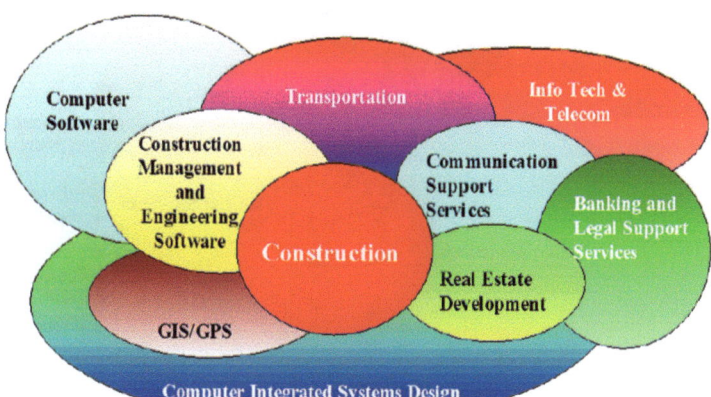

Fig. 19.10 Out of region construction and real estate development cluster

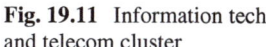

Fig. 19.11 Information tech and telecom cluster

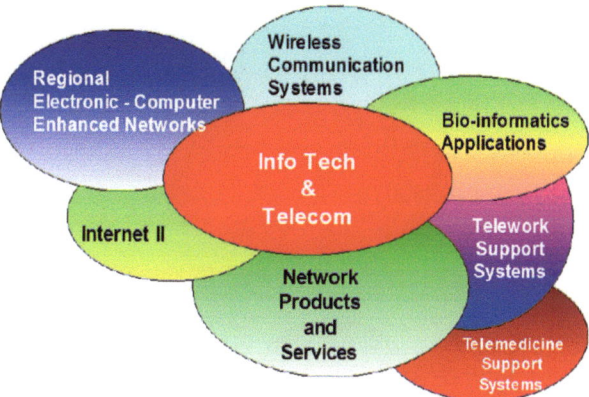

cluster. As such they identified several candidates for deepening (and strengthening) the cluster which included network design, network repair, network management, expanded network products and services, and training. Figure 19.19 presents candidates for deepening the tourism and finance clusters as well as for the IT&T cluster. Of particular interest here are the sub-components identified for the finance cluster (a quite weak cluster) including expanding securitization, electronic banking services and venture capital services. The deepening results for these three clusters are illustrated in Figs. 19.20, 19.21, and 19.22, respectively.

We now turn to the concept of cluster stretching as introduced in Fig. 19.23. Stretching the Tourism and Leisure Services cluster is illustrated in Fig. 19.24 where such options as electronic entertainment, attraction management and teleconferencing software, and electronically enhanced conferencing are presented (note that this exercise was conducted more than 15 years ago and thus the reason why the stretching options are not more advanced). Figure 19.25 illustrates the concept of

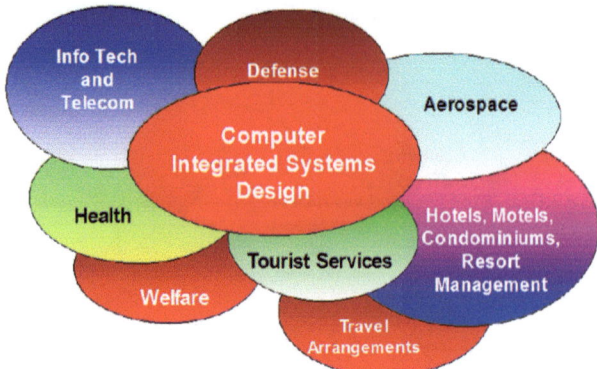

Fig. 19.12 Computer integrated systems design cluster

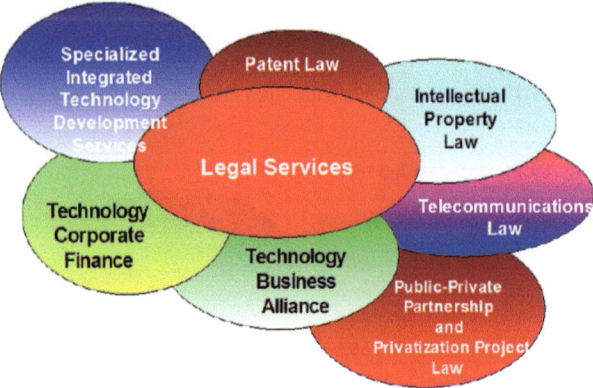

Fig. 19.13 Legal services cluster

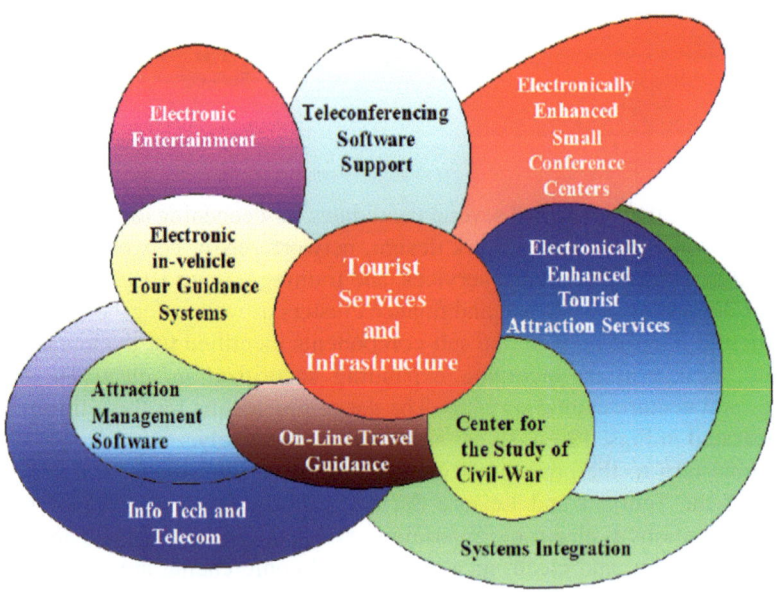

Fig. 19.14 Tourism cluster

Fig. 19.15 Finance cluster

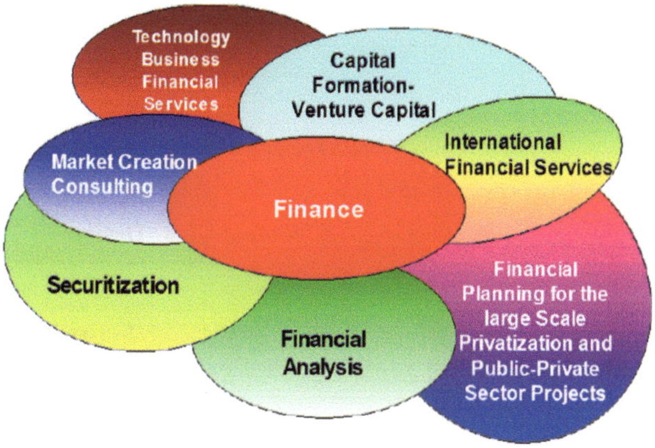

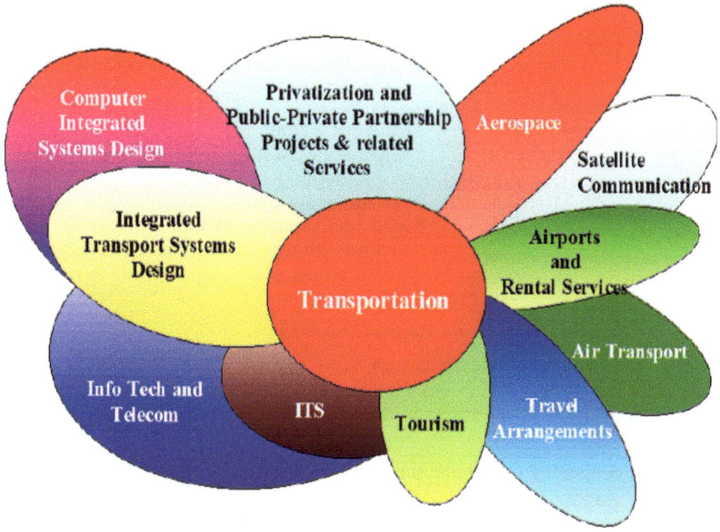

Fig. 19.16 Transportation cluster

Fig. 19.17 Competitive advantage

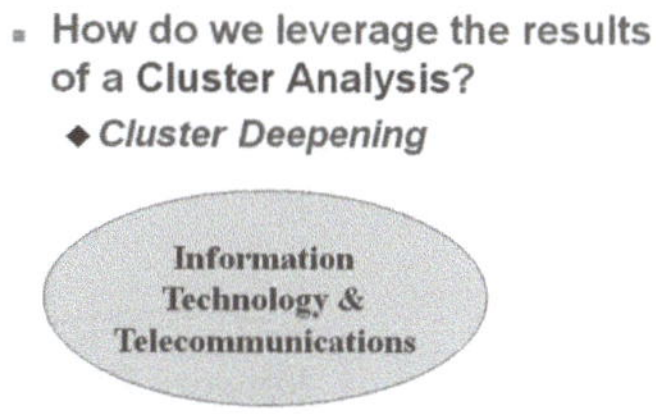

- How do we leverage the results of a **Cluster Analysis**?
  ◆ *Cluster Deepening*

inter-cluster linkage development. The process leading to the identification of these options involved presenting focus group participants with combinations of sectors/clusters and asking what new sectors or industry elements they thought might occur

Fig. 19.18 Cluster deepening

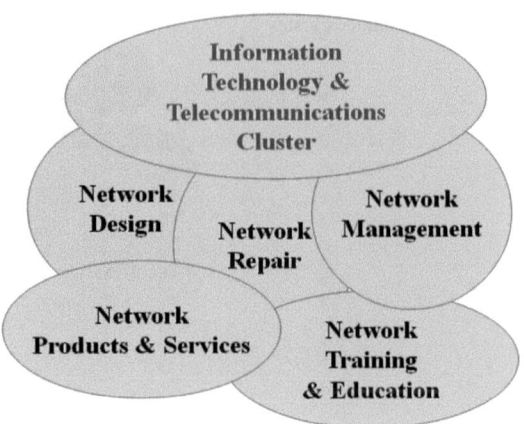

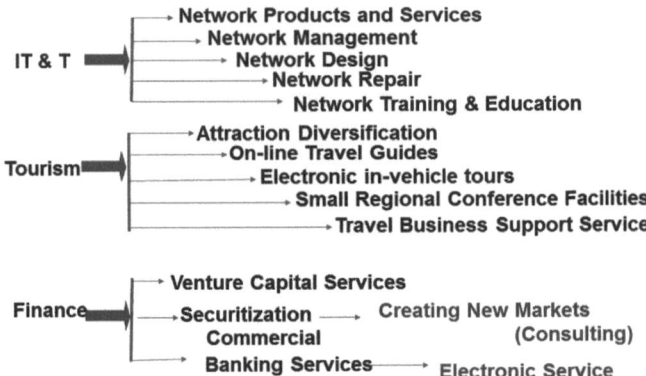

Fig. 19.19 Cluster deepening (specialization) development

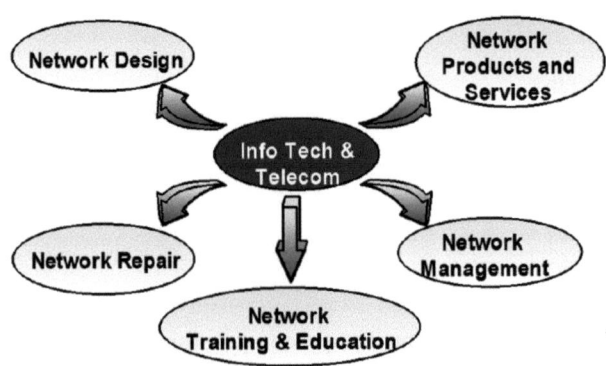

Fig. 19.20 Sectoral deepening (specialization) development: expanding securitization

Fig. 19.21 Sectoral deepening (specialization) development: electronic banking services

Fig. 19.22 Sectoral deepening (specialization) development: venture capital services

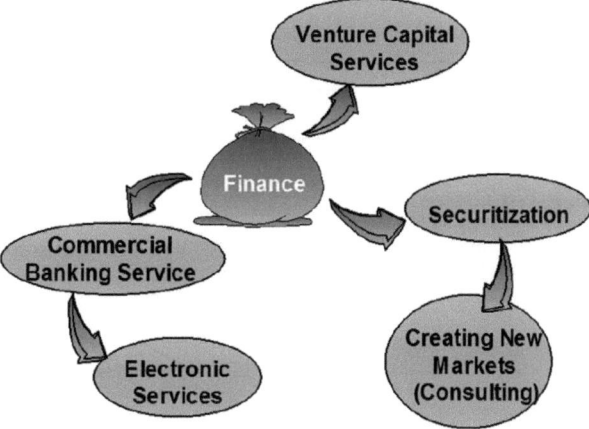

Fig. 19.23 Competitive advantage – 1

- How do we leverage the results of a **Cluster Analysis**?
  ◆ *Cluster Stretching*

at the interface. When posed with the question of what might be a future endeavor at the junction of the IT&T, system integration and transportation sectors, the participants identified vehicle route guidance systems and information services and amenities – all of which, it should be noted, are available in most new vehicles today some 15 years later.

The purpose of this part of the paper was to illustrate how cluster analysis and the front end rudiments for planning for their future development can be created and in particular with limited resources and time. While obtaining a valid description of the industry and its sectoral strengths, and thus the clusters of a region are important goals, finding fruitful directions for development are fundamental for directing

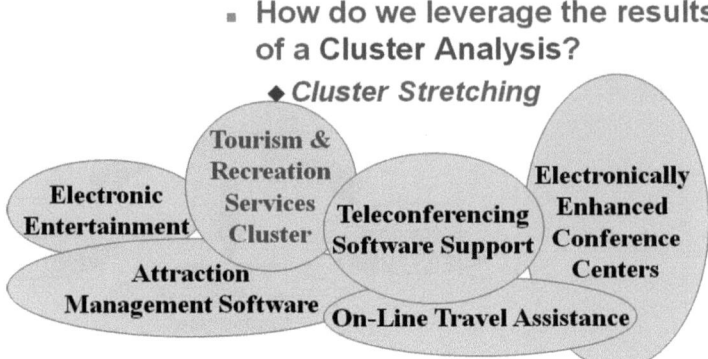

Fig. 19.24 Competitive advantage – 2

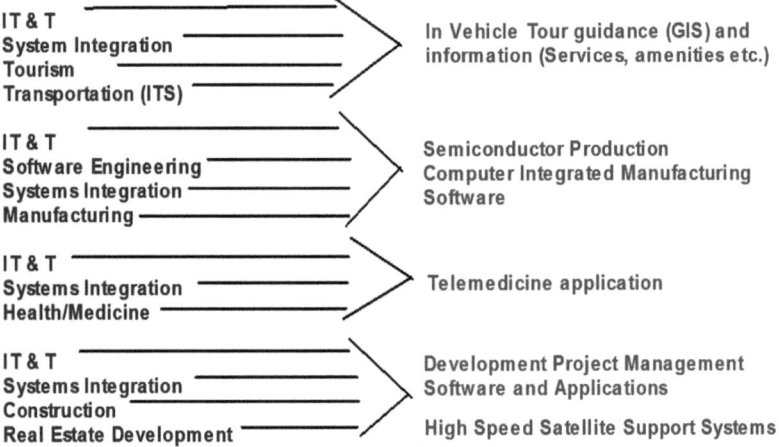

Fig. 19.25 Intersectoral linkage development

development planning and investment. The MSA and cluster deepening and stretching techniques are basic and effective tools for creating the road map for the future development of the region. We now switch to focus on another element of the Porter (1991) framework, related and supporting industries and organizations or the institutional infrastructure.

3.2 Supporting Industry and Cluster Organizations and Smart Infrastructure

One of the most important aspects of industrial clusters and their ability to support strong and dynamic economic growth and development is the institutional framework and support elements that exist (or do not) in a region. One way to think about this is the concept of smart infrastructure (Smilor and Wakeland 1990) i.e., the institutional infrastructure that is needed to support modern regional economic development. While Smilor and Wakeland's schema is somewhat dated in that some contemporary concepts such as knowledge and agglomeration are not mentioned there is still considerable value in there model for cluster led regional economic development planning and development.

There are four elements in the concept: talent, technology, capital and know-how (likely, we would call it knowledge and/or economically useful knowledge today). What is critical for the talent base of a region is of course its education and learning base, quality of life, creativity (Florida 2002) and the continuous configuring and re-configuring of its economic base in the high communication and innovation economies of the twenty-first century which all in turn impact the rate of immigration or emigration to or from a region.

Technology is a second component of smart infrastructure. The technical base of a region and its ability to renew and adapt to change depend on the availability of government (state and federal) resources for supporting technical development and promoting a supportive regulatory environment. The presence of strong and informed institutions to convert intellectual property into useful economic knowledge and to transfer that into commercially viable products is likewise of considerable importance in an era where the continuous innovation model of competition dominates. Also, science and technology research centers, science parks and business/technology incubation centers provide the base for high end innovations and their conversion into technological advanced businesses. Finally, systems for defining, protecting and marketing intellectual property, such as university technology transfer corporations and organizations are essential for growing dynamic and robust clusters and maintaining their contribution to regional economic development.

Capital is the third element making up a region's smart infrastructure. On the private side venture capital pools and tax advantages defined for such capital are important. R&D limited partnerships, small business investment corporations, and non-profits that operate capital networking events are increasingly recognized as part of the basic soft infrastructure of the region. At the same time there are technology venture support groups including angel investment funds, government programs supporting technology business ventures (SBIRs and the ATC program in the U.S.) and state and university venture capital funds. All of these are essential when advanced technology is a central element of a region's development base. Without such capital supporting functions a science and technology based cluster will gradually become obsolete and lose its competitiveness.

For the fourth element of smart infrastructure there are important business and technical support services that are essential. On the business side there is a need for incubation (specialized as well as broader based), education and training programs offered by universities and others, public and private sector organizations as well as professional support programs. On the technical side professional support organizations, capable local advisors and active business networks are a basic need. In short, diverse university college educational and training programs along with a mix of public and private organizations that ensure benefits from network events will enjoy maximum positive spillovers. All institutions and organizations that at base produce knowledge and know-how and knowledge spillovers are essential for the fourth major element of smart infrastructure.

Smart infrastructure is a foundation for successful economic growth and development and functioning and, because the twenty-first century is so knowledge and technology intensive, so is smart infrastructure. Much attention has been focused on the need for smart infrastructure because it is the foundation for smart cluster growth and development and the ability to adjust cluster activity to rapid changes in market preferences and technology. Within a specific cluster the presence of specialized institutions and organizations to promote the cluster and information among its constituents is of fundamental importance for survival and for re-inventing the cluster if and when it begins to decline. Now we give particular attention to three smart infrastructure examples in specific food clusters.

3.3 Examples of Smart Infrastructure in Food Clusters

This part of the paper includes three short synopses of food clusters in Europe. The purpose of presenting these is to illustrate the depth and breadth of smart infrastructure that supports them. The three are: the Oresund Food cluster in Denmark and Sweden; Greenport in South Holland and the Dutch Food Valley in the eastern part of the Netherlands in the lower Rhine River valley.

The Dutch Food Valley is centered on Wageningen as the "City of Life Sciences". As one would expect Wageningen University specializes in the life sciences and is ranked #2 globally in the area of food and nutrition research (http://www.food-info.net/wageningen.htm). The Food Valley is a food and nutrition cluster that is centered around and embedded in Wageningen University and a network of other surrounding specialized R&D organizations like NIZO which specializes in proteins and flavor/texture interactions and that also supports four university professors (Van Hoesel 2009). The university and related associated institutes employ 7,400 of which about 5,000 are senior researchers while the region also provides all levels of research and experimental facilities, incubators, a variety of public private partnerships, a well-integrated network of industry-university-technology transfer organizations, and linkages to capital providing organizations. The region claims that the Food valley is one of the largest food and nutrition clusters in the world (http://www.food-info.net/wageningen.htm). In short, it has a full complement of smart infrastructure.

The Food Valley is located in an agricultural heartland, the lower Rhine valley, and is centrally located on the major north-south population axis of European and in the midst of major highway, railroad and water transport routes. These location factors are core attributes as few food clusters around the world are found in places that do not have agricultural (food) production, access to transport infrastructure and to major markets. It also, as indicated above, has formed and grown an industry chain beyond food and nutrition research, and food production to include packaging, safety, production machinery, distribution, warehousing and quality control. In short, and again, it has all the locational and network infrastructure to support its continued growth and also an ability in its business culture to adjust rapidly to changing preferences and change in general thereby maintaining its market and market position.

The Oresund Food cluster in Denmark and Southern Sweden, as discussed above, was used to illustrate how changing global politics and policy in the form of the demise of the USSR and the rise of the European Common Market enabled it to greatly expand its footprint in Europe. We now provide a more detailed assessment (see http://www.oresundfood.org for source of the factual information below).

In the Oresund region in Denmark nearly 20 % of the labor force is employed in food industries, there are 4,000 scientists and food technologists and the region exports over 70 % of the food products produced. It is the first EU recognized center of excellence in food science. From the perspective of its population, which is one of the most highly educated in Europe, it provides a home test market as the region's consumers are well educated and "taste" demanding. The revenue turnover for the core food industry in the cluster's service area (Denmark and Southern Sweden) is nearly 100 billion Euro per year and employees in the food cluster have the highest value added in the food sector in Europe.

The Oresund region unlike the Wageningen Food Valley is a highly distributed cluster in that it includes parts of two nations (the Oresund region in Denmark and the Scania region in Southern Sweden). This in turn means that its components are highly distributed compared to the Food Valley. For example, the Oresund Food Network, an organization that represents the interests of the cluster boasts membership from 2 countries with more than 100 members including food manufacturers, bio-technology, ITC, innovation, professional associations, government agencies and universities. Because of the dispersed nature of the Oresund food cluster the R&D capability is embedded in multiple universities in both Denmark and Sweden as well as other science and research laboratories in both countries. Further, multiple research and innovation parks and incubators exist across the region to facilitate the conversion of knowledge into the commercial products. In short, the smart infrastructure has a dispersed network quality.

Again in the Oresund case location factors are at play in addition to historical precedent and political change (as noted above). Oresund is an agriculturally productive region and is located in close proximity to the core European market as well as having port access that enables it to maintain its large export market. Finally, the region is growing its smart infrastructure. While it does not appear that the elements of its smart infrastructure are as well developed as in the Food Valley, the highly dispersed geography and the fact that market access was greatly constrained due to political factors in Europe in the past, the region is now moving to rapidly consolidate and network together its more dispersed scientific, technology transfer, production and logistical infrastructure elements through the leadership of organizations like the Oresund Food Network

Greenport Holland is a different kind of food cluster in that its production is largely if not totally under glass, i.e., greenhouse production (see Van Vliet 2009 for supporting documentation for the facts cited below). It is located in South Holland and is in the most densely populated part of the Netherlands called the Randstad. The export value of its products and services is 15.7 billion Euros, 24 % of the production is horticulture and it commands 60 % of the world trade in flowers and plants and 90 % of the world trade in bulbs. Further, it adds an additional 2.5 billion in seed production. There are over 1,000 companies actively involved in the Greenport. Finally it is the oldest and largest greenhouse based cluster in the Netherlands.

The Greenport is well integrated into the national innovation platform in that flower and food production is one of the four key areas of the innovation strategy of the Netherlands. In the Greenport area a network of local enterprises, research institutes, local government and universities provide an integrated network for R&D and innovation. Recent and current projects include geothermal heating, construction of a Greenport in Shanghai, a geothermal power plant, and an electricity producing greenhouse. But the innovation and R&D work is even more divers and goes much beyond the locally focused work to include universities in many parts of the Netherlands, many science and research centers and parks, incubation programs that are all linked into the national innovation platform for agricultural products, agribusiness and horticulture.

The Greenport in Holland is nearby the country's main seaports in Rotterdam and Amsterdam, is located very near the food and flower auctions in the Netherlands, has immediate access to all forms of transport (land, water, rail and air), and being a part of the Randstad[4] means that it is centered amidst the most highly educated population in the Netherlands, a dense group of research institutes such as university research parks, TNO and the Technical University in Delft and which lie within one hour of the Wageningen Research University in the Food Valley. Finally, its location in the Rhine river delta means that it is at the terminus for the most populated river basin in Europe and thus has enormous access to the European market to go along with the global access provided by the transport system.

The three examples provide consistent stories about the need for smart infrastructure along with the more traditional location factors required to support a food cluster. High-end well networked and integrated R&D infrastructure coupled with application or knowledge converting infrastructure such as intellectual property services, incubators, science parks and extensive networks of business services advisors and organizations are in each case central to their global competitiveness. Further, integrated logistics elements are seen as central to each and all of the clusters' global competitiveness including packaging, production machinery, distribution, warehousing and quality control. Finally, the traditionally basic location factors are present in each of these cases: natural advantages: production base, market access and transport infrastructure. When this is coupled with high quality smart infrastructure food clusters as other clusters thrive.

---

[4] The Randstad is an intensely urbanized area enclosed in the triangle formed by Amsterdam, Rotterdam and Utrecht. The Randstad is home of the majority of the Netherlands population.

4 Conclusions

The paper began with an assessment of the relationship between economic development, entrepreneurship and industrial clusters noting that the evidence for a positive relationship between economic development and entrepreneurship was strong and with industrial clusters more ambiguous. Nonetheless, based on theoretical arguments, it was assumed that a positive relationship existed among all three. In short, it was assumed that they mutually reinforced each other. Further, it was also assumed that the importance of knowledge in the so-called "knowledge age" was very high. However, it was observed that while knowledge production and its related institutions are important it was recognized that the production of economically useful knowledge is even more important and that the entrepreneur is the central agent in converting knowledge into economically or commercially viable knowledge. Consequently, a significant part of the analysis in the paper focuses on the nature of these knowledge conversion facilitating institutions or infrastructure under the label of smart infrastructure which includes four dimensions: talent, technology, capital and know how. This then served as the foundation for the paper.

Next, the competitiveness of Northeast Asia for developing food clusters was examined and found to be quite supportive. However, the soft infrastructure that is critical to innovative cluster development and maintenance seems to be somewhat limited. This resulted on a focus on this issue in other parts of the paper.

Next the nature of cluster formation, growth and dynamics was addressed and used as the rationale for a quite detailed examination of the methods available to define and analyze clusters. Various techniques or tools were explained and their role in the analysis of clusters and for guiding the development of planning strategy for a region was illustrated. This was followed by a conclusion that the concept of facilitating or supporting institutions from the Porter (1991) cluster schema was of central importance for the growth and development of clusters in the twenty-first century when knowledge and its conversion into economically useful knowledge (commercial products) became a defining element of cluster competitiveness. Then the concept of smart infrastructure was introduced as a platform for examining this dimension and as a platform for examining the nature of smart infrastructure in three food cluster cases in Europe.

What are the conclusions that follow from the analysis? There are several. The review and presentation of various methods for identifying and nurturing clusters presented tools and techniques for conducting cluster analyses and for guiding subsequent planning for cluster strategy development. Among these the Multi-Sector Analysis (MSA) matrix is one of the most useful in that it enables a full assessment of strengths and weaknesses of potential cluster investment decisions. Methods for identifying clusters were also shown including input-output analysis and short cut methods that enable rapid development of results for planning purposes. Also, techniques for deepening and broadening cluster elements were demonstrated. The MSA matrix was used to illustrate how to identify critical bottlenecks for the growth and development of clusters in a region. But beyond these methods the concept of smart infrastructure was introduced to demonstrate the key elements of soft infrastructure and their integration as critical to the development and maintenance of competitive clusters. Finally, this conclusion was demonstrated with three food cluster examples from Europe.

A paper for presentation at the Northeast Asia Conference on Food Clusters in Niigata, Japan, January 23–24, 2010

References

Acs ZJ, Audretsch DB, Braunerhjelm P, Carlsson B (2004) The knowledge filter and entrepreneurship in endogenous growth. Center for Economic Policy Research, discussion paper, 4783

Acs ZJ, Audretsch DB, Braunerhjelm P, Carlsson B (2006) The knowledge spillover theory of entrepreneurship. Center for Economic Policy Research, discussion paper, 5326

Audretsch DB, Keilbach MC, Lehmann EE (2006) Entrepreneurship and economic growth. Oxford University Press, New York

Dent C (2008) East Asian regionalism. Taylor & Francis, London

Florida R (2002) The flight of the creative class: and how it's transforming work, leisure and everyday life. Basic Books, New York

Hart DM (ed) (2003) The emergence of entrepreneurship policy: governance, start-ups, and growth in the U.S. knowledge economy. Cambridge University Press, Cambridge

Hewings GJD (1985) Regional input-output analysis. Sage Publications, Beverly Hills

Information Design Associates and ICF/Kaiser International Inc. (1997) Cluster-based economic development: a key to regional competitiveness. Report to the US Department of Commerce, Washington, DC

Krugman P (1995) Development, geography and economic theory. MIT Press, Cambridge, MA

Lagnevik M, Sjoholm I, Lareke A, Ostberg J (eds) (2003) The dynamics of innovation clusters: a case study of the food industry. Edward Elgar, Cheltenham

Marshall A (1920) Principles of economics, 8th edn. Macmillan, London

Perroux F (1950) Economic space: theory and application. J Econ 4:90–97

Porter ME (1991) The competitiveness advantage of Massachusetts. Harvard Business School, Boston

Rocha HO (2004) Entrepreneurship and development: the role of clusters. Small Bus Econ 23:363–400

Smilor RW, Wakelin M (1990) Smart infrastructure and economic development: the role of technology in global networks. In: Kosmetsky G, Smilor RW (eds) The technopolis phenomenon. IC2 Institute, University of Texas, Austin, pp 53–75

Stimson R, Stough RR, Roberts R (2006) Regional economic development: analysis and planning strategy. Springer, Heidelberg

Stough RR, Haynes KE, Campbell HS Jr (1997) Small business entrepreneurship in the high technology services sector: an assessment for the edge cities of the U.S. national capital region. Small Bus Econ 9:1–14

Van Hoesel R (2009) Food valley Wageningen. Presentation at technopolicy annual conference on shaping science based clusters: creating a joint regional agenda, Wageningen, the Netherlands, 24–25 Sept 2009. www.technopolicy.net

Van Vliet E (2009) Greenport Westland-Oostand. Presentation at technopolicy annual conference on shaping science based clusters: creating a joint regional agenda, Wageningen, the Netherlands, 24–25 Sept 2009. www.technopoligy.net

MIX
Papier aus verantwortungsvollen Quellen
Paper from responsible sources
FSC® C105338

If you have any concerns about our products,
you can contact us on
ProductSafety@springernature.com

In case Publisher is established outside the EU,
the EU authorized representative is:
**Springer Nature Customer Service Center GmbH
Europaplatz 3, 69115 Heidelberg, Germany**

Printed by Libri Plureos GmbH
in Hamburg, Germany